Flood Risk in a Climate Change Context - Exploring Current and Emerging Drivers

Edited by Tiago Miguel Ferreira and Haiyun Shi

Published in London, United Kingdom

Flood Risk in a Climate Change Context - Exploring Current and Emerging Drivers
http://dx.doi.org/10.5772/intechopen.100704
Edited by Tiago Miguel Ferreira and Haiyun Shi

Contributors
Shirley Gato-Trinidad, Jamie Carroll, Ezreena Aladin, Tristan Gilbert, Johnny Douvinet, Haiyun Shi, Yao Wang, Suning Liu, Zhaoqiang Zhou, Trond Knapp Haraldsen, Margherita Rago, Mayra Alejandra Estrella Nuñez, Alkmini Firtinidou-Stergiou, Chenxin Jonathan Yee, Alberto Barontini, Tiago Miguel Ferreira, Daniel V. Oliveira, Yasumitsu Maejima

First published in London, United Kingdom, 2023 by IntechOpen
IntechOpen is the global imprint of INTECHOPEN LIMITED, registered in England and Wales, registration number: 11086078, 5 Princes Gate Court, London, SW7 2QJ, United Kingdom

British Library Cataloguing-in-Publication Data
A catalogue record for this book is available from the British Library

Additional hard and PDF copies can be obtained from orders@intechopen.com

Flood Risk in a Climate Change Context - Exploring Current and Emerging Drivers
Edited by Tiago Miguel Ferreira and Haiyun Shi
p. cm.
Print ISBN 978-1-80356-602-3
Online ISBN 978-1-80356-603-0
eBook (PDF) ISBN 978-1-80356-604-7

An electronic version of this book is freely available, thanks to the partial funding support of libraries working with Knowledge Unlatched. KU is a collaborative initiative designed to make high quality books Open Access for the public good. More information about the initiative and links to the Open Access version can be found at www.knowledgeunlatched.org

Meet the editors

Dr. Tiago Miguel Ferreira is a lecturer in Civil Engineering at the University of the West of England (UWE Bristol), UK, and an invited assistant professor at the University of Coimbra, Portugal. Dr. Ferreira's academic activity has been dedicated to studying the structural vulnerability of historical buildings and urban areas to natural and anthropogenic hazards, particularly earthquakes, fires, floods, and landslides. Ranked twice among the top 2% of scientists in the world by Elsevier BV and Stanford University, he is the author of more than 200 scientific and technical publications in his research fields. Dr. Ferreira has also maintained an intense editorial activity over the last few years, currently holding the position of editor-in-chief for two international journals.

Dr. Haiyun Shi is an associate professor at the Southern University of Science and Technology, Shenzhen, China. He obtained his Ph.D. from Tsinghua University, Beijing, China. He previously worked as a postdoctoral fellow/senior research assistant at the University of Hong Kong. His research interests include digital watersheds, hydroinformatics, climate change, hydrological extremes (floods and droughts), and sustainable development. He has been invited as a reviewer for more than thirty international journals and was an expert reviewer of the "IPCC Sixth Assessment Report." He has also served as an editorial board member and guest editor of six international journals. Dr. Shi has served as the main convener of two sessions at the Asia Oceania Geosciences Society annual meeting and the main convener of one session at the American Geophysical Union Fall Meeting.

Contents

Preface

In this timely and comprehensive edited volume, the reader embarks on a journey through the complex and multifaceted aspects of floods and their intricate relationship with climate change.

Floods have long been recognised as one of the most significant natural hazards, posing substantial threats to communities and ecosystems worldwide. However, in the context of a changing climate, these risks are further amplified, necessitating a deeper understanding of both existing and emerging risk drivers. With that in mind, this volume focuses on the dynamic nature of floods and their intricate relationship with climate change.

Through a journey across chapters, each focusing on a specific aspect of floods and their risk drivers, this volume sheds light on the multiple factors influencing flood occurrence, intensity, and impacts in our rapidly changing world. From exploring the role of constructed wetlands in stormwater management to assessing flood risk in historic urban areas, analysing non-stationarity of extreme rainfall to advancing torrential rainfall forecasts, and examining flood damage on agricultural land to understanding post-fire debris flow susceptibility, every facet is meticulously investigated here to obtain knowledge and actionable insights.

Chapter 1 focuses on the role of constructed wetlands as a nature-based solution for stormwater management. Integrating nature into urban landscapes, constructed wetlands offer a sustainable approach to managing stormwater and reducing flood risks. This sets the stage for the exploration of effective flood management strategies.

Chapter 2 takes the reader to the historic city centre of Aveiro in Portugal. Through a large-scale flood risk assessment, a multidisciplinary, international research team applies a methodology to identify buildings and areas of Aveiro at risk. Based on this assessment, the authors provide essential insights into the flood risk scenario and possible strategic rehabilitation interventions for protecting this area while preserving the heritage value of its buildings.

In Chapter 3, the focus of the reader shifts to the middle and lower reaches of the Yangtze River basin in China. By analysing the non-stationary characteristics of extreme rainfall events, the authors explore the correlations of extreme rainfall with large-scale atmospheric circulations such as El Niño–Southern Oscillation and Western Pacific Subtropical High. The discussion and results offered in this chapter demonstrate how accounting for climate change-induced shifts in flood magnitudes and frequencies can lead to better-informed decision-making and, consequently, to more resilient infrastructure.

Advancements in torrential rainfall forecasts and their role in improving early warning systems take centre stage in Chapter 4. In this chapter, the author explores the use of innovative approaches and technologies, such as high-resolution weather models and ensemble forecasting, to enhance the accuracy and lead time of torrential rainfall predictions. As shown in this chapter, such tools and applications can play a critical role in empowering communities to take timely and appropriate actions in response to potential flood events.

Chapter 5 explores the impact of catastrophic floods on agricultural land in cold regions with glaciers. By examining examples of the impacts of floods of different sizes in these areas, the authors investigate the effects of floods on yields and possible measures aimed at reconstructing soils with the ultimate goal of regaining productivity. With this discussion, the author provides insights into mitigating the long-term effects of floods on agricultural systems and highlights the importance of safeguarding food production and rural livelihoods.

Finally, Chapter 6 focuses on the increasingly relevant topic of post-fire debris flow events. In this study, the authors apply the cellular automaton RuiCells to assess flash flood susceptibility in sedimentary areas affected by extreme fires. The findings of this study highlight the relationships between morphological patterns, surface water flows, and debris flow susceptibility and showcase the importance of understanding the relationship between wildfires and flood risks for the definition and implementation of effective proactive measures aimed at enhancing resilience and protecting communities in fire-affected regions.

Together, these six chapters offer valuable insights and innovative approaches. By embracing a multidisciplinary perspective and unravelling the complex interactions between climate change, natural systems, and human activities, this volume equips researchers, practitioners, and policymakers with the knowledge and tools necessary to prepare for and mitigate the impacts of floods in a more effective way, with the goal of contributing to a more resilient and sustainable future where communities worldwide are better prepared to face the challenges posed by floods in a climate-changing context.

Dr. Tiago Miguel Ferreira
School of Engineering,
Department of Geography and Environmental Management,
University of the West of England (UWE Bristol),
Bristol, United Kingdom

Dr. Haiyun Shi
School of Environmental Science and Engineering,
Southern University of Science and Technology,
Shenzhen, China

Chapter 1

Understanding the Role of Constructed Wetlands in Stormwater Management

Shirley Gato-Trinidad, Jamie Carroll, Ezreena Aladin and Tristan Gilbert

Abstract

Constructed wetlands have been utilized for some time in the treatment of wastewater and have been recognized for the treatment of stormwater runoff and flood protection in the last couple of decades. Constructed wetlands are built to remove sediment and nutrients, primarily phosphorus and nitrogen, from contaminated water. However, with increased urbanization and enhanced climate change, these constructed wetlands need to be managed and their treatment effectiveness monitored and maintained especially at the post-construction phase. In addition, a greater understanding of the role of these systems in the urbanized environment and how they treat wastewater are needed to optimize their performance. As more advanced computer modeling is developed there is a need to ascertain what parameters and how these changes overtime and what skills are required to enable the adoption of constructed wetlands for future planning and management. There has been limited research into constructed wetlands for flood mitigation and with some receiving inflows larger than their design intent, it is necessary to determine if these systems would still be able to treat pollutants. This chapter involves a review of the literature to address these concerns relating to constructed wetlands.

Keywords: constructed wetlands, pollutants reduction, flood mitigation, MUSIC modeling, TN, TP, TSS

1. Introduction

Whilst constructed wetlands have been utilized for some time in the treatment of wastewater, they only gained popularity for the treatment of stormwater runoff and flood protection in the last couple of decades [1, 2]. Constructed wetlands are employed to remove sediment and nutrients, primarily phosphorus and nitrogen, from contaminated water [1]. However, with increased urbanization and enhanced climate change, these constructed wetlands need to be managed and their treatment effectiveness monitored and maintained once these are established.

Constructed wetlands remove sediments using large ponds which allow for dissipation of water velocities, making sediment particles drop out of the water column, settling at the bottom of the basin which is then supposed to be cleaned regularly [1]. The removal of nutrients by constructed wetlands is varied and complex and is understood to be primarily undertaken by anoxic microbial degradation processes within the wetland environment [3]. Secondary to this, plants also capture nitrogen and remove it from the system through various methods such as volatilization and phytodegradation [1]. Wetland performance in treating stormwater is generally a function of hydraulic loading rate and detention time and these two parameters are in turn functions of the runoff volume, storm intensity and the wetland size itself [4]. Sizing the wetland is crucial for the health of the wetland treatment system. The hydrodynamic criteria of wetlands such as the inundation depth, wetness gradient, base flow and hydraulic regime are crucial for wetland sizing. If these hydrodynamic characteristics receive inadequate attention, the performance of treating stormwater is likely to be reduced.

Constructed wetlands also provide a retarding function, and therefore can be utilized to assist in flood protection in urban areas. As wetlands are generally controlled by a pit and a piped outlet, they can act under the same principle as a retarding basin by discharging flood flows at a controlled rate. With more wetlands being constructed and restored widely, flood storage capacity is increased, and the flood peak is reduced. Constructed wetlands and the restoration of wetlands have come into wide practice as they have the potential to act as an effective water treatment basin as well as provide essential flood control [5].

In Australia, particularly in metropolitan areas, constructed wetlands have become a common occurrence [3]. There is a multitude of reasons why wetlands have increased in popularity over the years, some reasons include their effectiveness in treating stormwater, their ability to treat large areas, scalability, their cost-effectiveness, ease of maintenance, and because they can act as a feature [6, 7]. Perhaps the main driver for their increase in popularity in Victoria, Australia is the legislation that requires all new developments to treat the additional runoff caused by the newly installed impervious surfaces to acceptable levels so that the concentrations of nutrients in receiving waters are not compromised [8].

The Best Practice Guidelines recommend the removal of 80% TSS (Total suspended solids) [9]. This is generally managed practically by sizing a constructed wetland to treat 100% of rainfall flows up to 1 in 3 months ARI rainfall intensity, which in Victoria represents on average 80% of the total annual rainfall [10]. Depending on the design, i.e., if the wetlands are constructed "online" as opposed to "offline", constructed wetlands may experience inflows from larger rainfall intensities. However, the best practice guidelines do not consider detention characteristics of the wetlands.

To date, there has been limited research into constructed wetlands treating flows larger than its design intent. Whilst there have been studies which investigate the capabilities of constructed wetlands over long periods, e.g., 2 years, which cover a range of storms [11, 12], there appears to be little research directly focusing on the treatment that occurs from wetlands receiving flows larger than their design capacities.

2. Why are constructed wetlands utilized in the urban environment?

Urban Stormwater: Best Practice Environmental Management (BPEM) Guidelines [9] outline the principles and objectives behind the use of water treatment devices such as wetlands by referencing SEPPs (State Environmental Protection Policies) of Victoria, Australia. These objectives are clear: to preserve the beneficial uses of local waterways, including:

- natural aquatic ecosystems and associated wildlife;
- water-based recreation;
- agricultural water supply;
- potable water supply;
- production of mollusks for human consumption;
- commercial and recreational use of edible fish and crustacea; and
- industrial water use.

As urbanization increases, so too does the anthropogenic environmental impact, which includes pollutants occurring from human lifestyle activities, building and infrastructure, construction activities [9], urban vehicular emissions and increased stormwater runoff volume peaks and nutrient loads due to hard surface runoffs inherent in typical urban development [13]. Constructed wetlands, as part of a SUDS (Sensitive Urban Drainage System) can be used to mitigate these impacts [14], and within Victoria, Australia they are typically installed as part of large-scale urban developments to comply with the BPEM guidelines.

Wetlands are cost-effective treatment systems that can be used to treat urban stormwater runoff. The advantages for installation of constructed wetlands, including "(1) low cost of construction, and, especially, maintenance; (2) low energy requirements; (3) being a 'low-technology' system, they can be established and run by relatively untrained personnel; and (4) the systems are usually more flexible and less susceptible to variations in loading rate than conventional treatment systems" [3]. In addition, "the major disadvantages of constructed wetland treatment systems are the increased land area required, compared to conventional systems, and the possible decreased performance during winter in temperate regions" [3]. Under Melbourne Water's drainage Schemes, landowners and developers are fairly compensated for land use dedicated to wetlands, thus mitigating the disadvantage of the increased land area required [15]. As wetlands are integrated into large-scale urban development design and construction, they are typically designed to receive flows up to the 1 in 3-month ARI rainfall to minimize land take whilst treating to best practice.

3. How do constructed wetlands treat stormwater?

To enable efficient operation and maintenance of constructed wetlands, the current practice is to split the treatment processes into separate zones; the sediment pond and macrophyte zone. **Figure 1** shows the design layout for constructed wetlands within Melbourne as per the guideline and **Figure 2** indicates the sediment storage pond.

3.1 Sediment removal

The sediment pond is utilized to remove sediments before the wetland. Current design guidelines require that sediment ponds shall be sized to retain 95% of sand particles (i.e., particles down to 125 μm) during a 1 in 3-month ARI rainfall event [16].

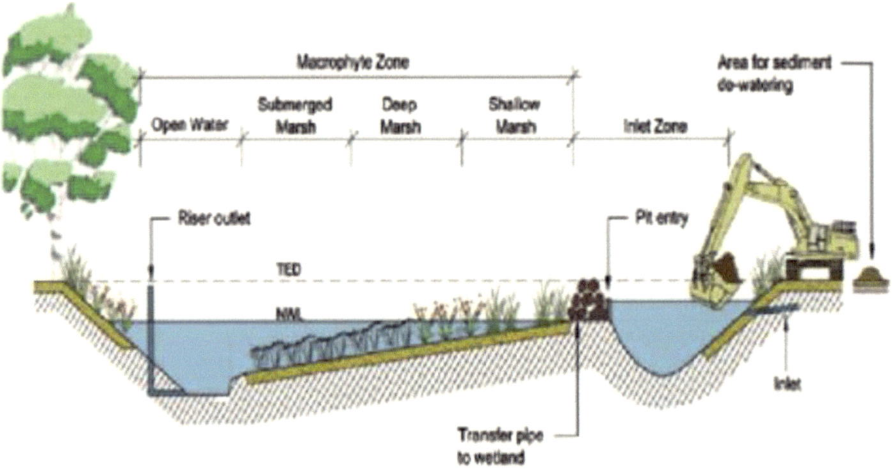

Figure 1.
Constructed wetland diagram [16].

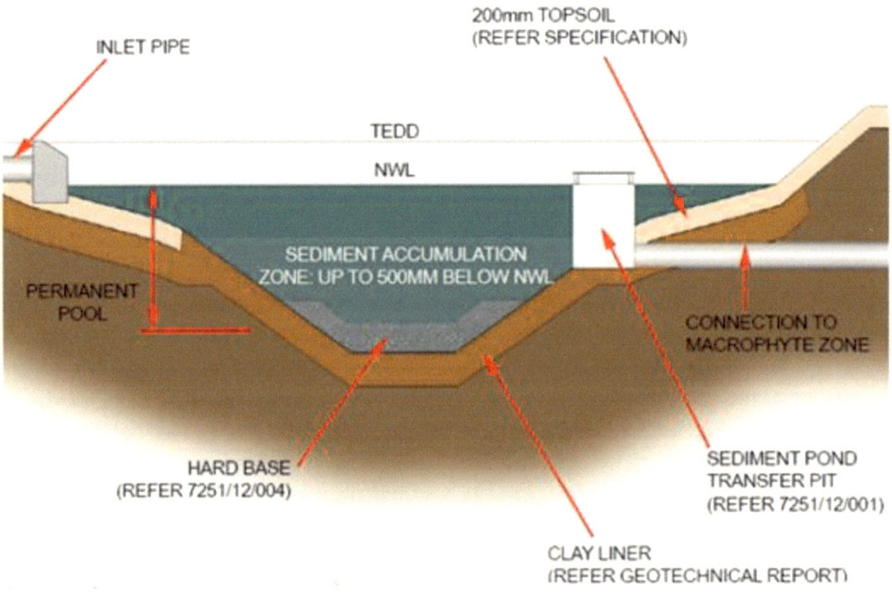

Figure 2.
Sediment pond storage [17].

Regular cleanout of retained sediment is required whenever retained sediment levels reach 500 mm below the normal water level (NWL) [17]. There is conjecture as to how well this maintenance is carried out, as many existing constructed wetlands do not receive regular sediment maintenance.

Secondly, the Macrophyte Zone (shallow water-filled ponds planted with aquatic plants) is utilized to contain the remaining sediments not captured by the sediment pond by the velocity reducing and filtering effect of Macrophyte (aquatic plant) stems and root systems [18].

3.1.1 Nitrogen removal

The macrophyte zone removes nitrogen through the Nitrification–denitrification process [19, 20]. Sediment ponds also provide limited nutrient treatment. This includes an attachment of approx. 15% of Total Nitrogen (TN) and Total Phosphorus (TP) to particles larger than 300 um [21], and small amounts via anoxic biochemical processes [3]. Studies on the Prado wetlands in California USA suggest that constructed wetlands have, on average, 50–60% TN removal [22].

3.1.2 Phosphorous removal

Phosphorus removal is primarily through sorption by clay, and typically will reach a saturation point after which phosphorus can leach back into the water it should be treating [23], and upon reaching this point, wetland clay materials should be replaced. Additional sorption can be achieved via decomposed plant matter [23]. Studies on the Prado wetlands in California USA suggest that constructed wetlands have, on average, 40–50% TP removal [22].

3.2 On-line/offline wetland operation

Standard wetland designs in Victoria include two typical layouts: online wetlands with the sediment pond receiving unrestricted flows and the macrophyte zone receiving controlled flows, and alternatively, offline wetlands where flows into the wetland can be restricted using a diversion pit [16]. Online flows include the total runoff from a catchment area. Offline flows divert a volume of the flow up to a certain amount, and the remaining flow bypasses the wetland. If a wetland does not include a sediment pond, then the bypass can be managed via a pit arrangement with suitable baffle, invert levels and pipe sizes to suit the urban catchment area.

Wetlands that are receiving an online flow passing through the Macrophyte zone are at risk of having collected sediments and nutrients dislodged and washed

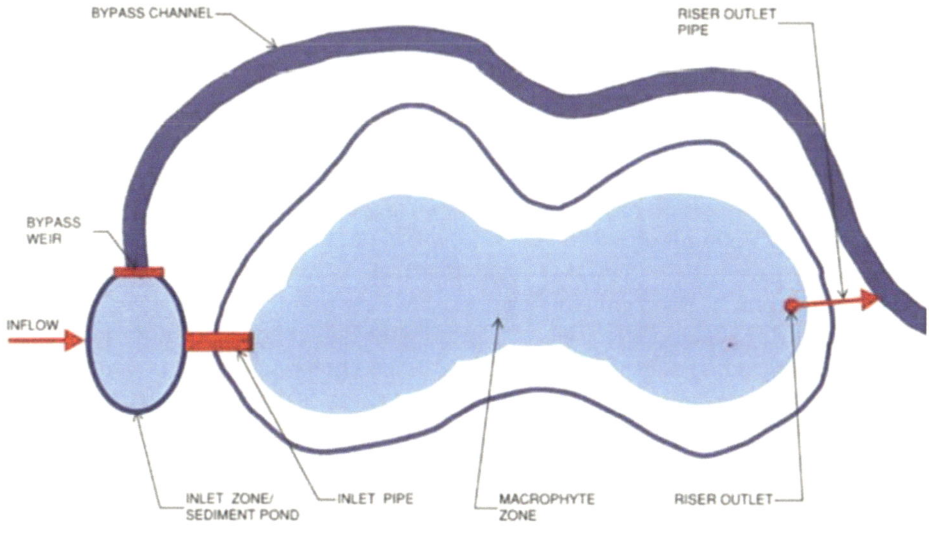

Figure 3.
Schematic diagram of a constructed wetland, adapted from [16].

downstream during peak flows [16]. Standard wetland design practice includes a bypass channel to prevent this occurrence as shown in **Figure 3**.

4. How do wetlands provide flood control?

Severe flood events can be distinguished as a natural disaster as its effect includes damages to properties and agricultural lands and in some cases loss of life. Flooding may be caused by dam failures, snow melts and when a large amount of rainfall occurs, and the natural waterways do not have enough capacity to convey excess water and result in overland flow. Overland flow can result from two hydrological processes: the first process is through a big storm event, where the rainfall intensity is large and it cannot infiltrate into the soil and the other process is when the soil is oversaturated - where there is no more capacity for the soil to hold extra rainwater [5]. Rain that falls onto the surface can either go through the evaporation process, get infiltrated into the soil, run along the impervious surfaces or get captured in hollows surface of the ground or wetlands. In the last 150 years, we have lost almost 70% of the capacity of the soil to hold water due to developments being built and more impervious areas installed [24]. This capacity needs to be restored to reduce the amount of overland flows and the risk of properties and people getting inundated. To minimize the risk of flooding and to protect assets and properties downstream, engineers have manipulated the use of constructed wetlands for flood mitigation and control.

In the past, engineers used structural restraints such as levees, which is not always the optimal solution to prevent flooding. Unfortunately, levees tend to hold up water which can significantly increase the level of the river stage and increase the flood velocity. There are a few flood cases where flood water would surge and overtops levees like the flood event in 1993 along the Mississippi River and Missouri River [24]. The flooding in mid-July 1993 from the Mississippi River and Missouri River exceeded the 100-year average recurrence interval (ARI) and had caused major properties damage cost of between US$ 12 billion–US$ 16 billion and approximately 32 losses of life [25]. As more areas are being inundated by overland flow and more properties are damaged, engineers have re-evaluated the situation and have considered other alternatives for flood mitigation plans and developing effective floodplain management programs. The aftermath of the great flooding of the Mississippi and Missouri River, wetlands have been considered as part of the flood management and flood mitigation process [26]. Wetlands have been installed and restored in many areas by the US federal government and the US Army Corps of Engineers in the United States and flooding at the downstream of the rivers was reported to have been reduced [27].

Constructed wetlands and the restoration of wetlands have come into wide practice as they have the potential to act as an effective water treatment basin as well as providing essential flood control. Wetland can alter flooding in many ways such as reducing the peak flood water level, the timing of flood water can be delayed, or the flows can be reduced by providing flood storage [5]. The location of wetlands is important for implementing flood protection. Wetlands that are located at an upstream location, the wetland will mainly be affected by headwater from rainfall, whereas wetlands that are located at the downstream location will mostly be affected by river flow. Constructed wetland at the downstream locations is often dry before a storm event therefore it has the potential to store more water during a storm event [5].

The sizing of wetlands is also equally important to the location of wetlands where continuous hydrologic modeling must be undertaken to simulate wetland storage during major storm events. Moreover, the time of concentration that a large flood needs to reach the wetlands must be taken into consideration. With the right computed concentration time, it can help reduce flooding at the downstream location just before the flood peaks [26]. Like a retarding basin, flood water that is stored in wetlands helps to delay the peak time of downstream hydrograph by releasing its water slowly and in a controlled manner. According to [26], the drainage of wetlands can influence flood levels where the storage of water attenuates and potentially delay downstream flood peaks.

Constructed wetlands are designed to replicate natural wetlands such as meadows, saltwater marshes, forested wetlands, and bogs. Aquatic or wetlands plants are used in constructed wetlands not only to reduce the amount of pollutants for stormwater quality but also to offer an ecological habitat to a wide range of wildlife species. Moreover, wetlands plants can create great landscape features and recreational amenities for the community [28]. Wetlands not only act as a flood storage but also have the capability to reduce the velocity of flood water with the influence of wetlands plants or vegetation. According to [29], velocities of flood water through wetlands are usually delayed by a friction factor which is influenced either by the slope, depth and/or the vegetation type (density and height of plants). This friction factor also known as the Manning's roughness coefficient developed by Chow in 1959 is widely applied by engineers to calculate the resistance of flow in open channels.

Velocity rate can be altered depending on the surface roughness where a higher Manning's value can significantly reduce the velocity of water. This can be seen in Manning's equation where velocity is respectively proportional to the roughness coefficient [29]. Wetland's riparian or vegetation have higher Manning's value ("n") in comparison with "n" value of a concrete path (Vegetation, n = 0.06 and concrete path, n = 0.02). Therefore, the vegetation of wetlands can influence flows in which the roughness of wetlands channel bed can potentially reduce the velocity of flood water and reduce the peak flow discharge [5]. According to [30], with the presence of wetland plants, the velocity profile is uniform in the vertical direction. This is illustrated in **Figure 4**. The velocity of flows through wetlands may be reduced significantly depending on the vegetation density and the height of the wetland plants increase friction and drag by the vegetative stems [30].

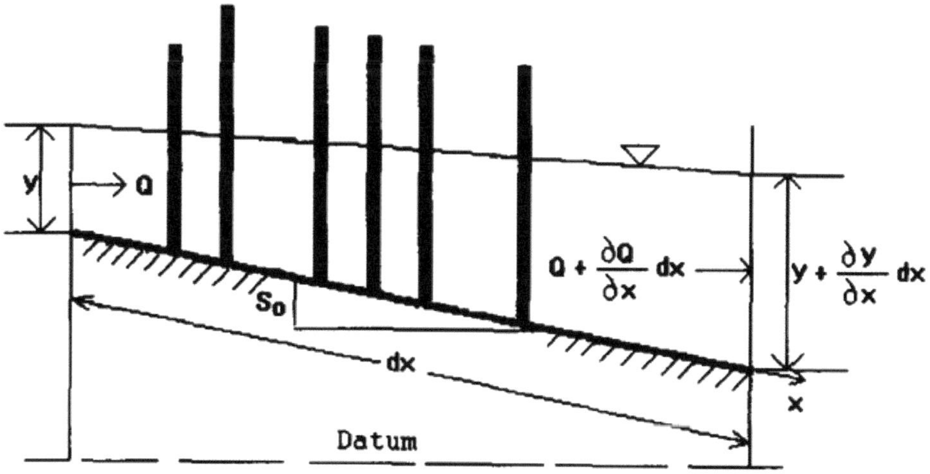

Figure 4.
Schematic elevation view of a wetland emergent plant (adapted from [30]).

The importance of wetlands for flood mitigation is now recognized and understood widely. Ogawa and Male [31] performed a hydraulic simulation on evaluating the flood mitigation potential of wetlands for the "Charles River, Neponset River and Ten Mile River in Massachusetts". From the simulation, the results implied that both upstream and downstream wetland locations altered peak flows and reduced flooding. Another study of wetland's role for flood mitigation was done for the Red River Valley major flood in 1997 and the damage cost by the flood that year was US$ 3.5 billion. The study concluded that a 5% increase in wetland area would significantly reduce flood volume by 5.6% for the 1997 flood event and also reduce the amount of damage cost [32]. Moreover, the restoration of wetlands within the Devils Lake basin of North Dakota could potentially store 72% for the 2 Year ARI storm event and 41% for the 100 Year ARI storm event of total runoff and the US Corps of Engineers calculated that flood damage cost (Approximately US$ 17 million) can be prevented each year with the use of wetlands for flood mitigation [5]. Wetlands are not only used to treat stormwater but also play an important part in reducing flood peaks and flooding at the downstream locations.

5. Are existing wetlands capable of treating flows larger than what they are designed for?

The most crucial part of designing wetlands is the sizing of wetlands. It is recommended that the capacity of a wetland should be at least 3% of its catchment size or can take 1 in 3 months flow to remove pollutants within the guidelines for stormwater quality treatment [17]. Since there is limited research on wetlands receiving flows larger than their design intent, this section will review the following fundamental questions:

5.1 How do depth, duration, and frequency of flooding influence wetlands plants?

The duration and frequency of overland flow (water regime) can be a major influence on the development of wetland plants. In Australia, wet to dry seasonal changes cause different water levels each season which can potentially affect plant growth and responses [33]. Wetlands plants are utilized to prevent erosion, capture fine particles and to trap pollutants from runoff [3, 17].

An experiment was done to determine whether the depth, duration and frequency of flooding influence the development of wetlands plants. The experiment was done with different combinations of depth, duration, and frequency of flooding with different types of wetlands plants collected from (a) wetlands that are rarely dry (near permanent) and (b) poorly drained wetlands (intermittent) in New South Wales [33]. All 17 treatments were placed outdoor in uncovered tanks to allow the plants to be exposed to rainfall and some tanks were filled with water depending on the treatment. According to [33], the experiment was conducted for more than 16 weeks using different combinations to maximize results as plants are well established after this period of time. The results show that the depth of flooding does have a significant impact on the biomass and the different species of wetland plants for both wetland types (near permanent and intermittent) and the results of the durations of inundated plants varied between plants from the two wetlands.

The results indicated that the inundated plants from the near-permanent wetlands show no difference between different durations of flooding (4, 8 and 12 weeks) whereas plants from the intermittent wetlands show there was a decline in some species of plants. As for the flooding frequency, no significant impact on plants establishments for all 17 treatments for plants that are from the near-permanent wetland but for the plants from the intermittent wetlands some decline of plants species are reported. The highest biomass and the vast plant species can be seen from the treatment tank that were never flooded [33]. According to [34] some wetland plants species are sensitive to the change of water level and could cause a distinct loss in its species which are caused by oxygen depletion and the ability to go through the photosynthesis process when it is fully submerged. Therefore, depending on the wetland type (near permanent and intermittent) wetlands plants that are frequently flooded can have an adverse effect on the treatment process.

5.2 Does the residence time distribution get affected with the change of wetland depth and flow rate?

According to [35], "A distribution of times that parcels of water spend in a constructed wetland is known as a residence time distribution". The residence time distribution (RTD) is a tool that has been used widely by engineers to measure wetland's characteristic that could affect its treatment capabilities. Retention or residence time can be calculated as retention time = (area x depth)/flow per day where flow is the average inflow and outflow of the wetland noting that the velocity profile in subsurface flow wetlands can influence the distribution of the hydraulic residence time. This is due to the velocity profiles that are influenced by the surface friction that potentially slows the water [36]. The RTD can be modeled on a computer and using tracer tests. A tracer test involves a dyed chemical tracer that gets dissolved into the wetland which can easily be detected to measure RTD [35]. Holland et al., [36] has conducted a study on RTD characteristics and how wetland depth and flow rate can influence it. The study was performed by using a series of dye tracers in a small-scale constructed wetland with a constant and controlled flow rates and water levels. Different water levels and flow rates are then used throughout the study to determine its effects on RTD.

Twelve experiments were conducted with different sets of water levels and flow rates, over a period of 13 weeks by adding 15 l of dye to the inlet of the constructed wetland. The average depth of water is 166 and 398 mm for low and high-water levels respectively. The results of this study indicated that there was no significant difference for the RTD values between the high and low flow rates (**Table 1**). There was however a significant difference for the RTD between the high and low water depth, the low water level result in distribution with one clear peak value whereas the higher water level result in a continuous probability with two different peaks. According to [36], the value differences are reflected in **Table 1** for the water levels are significant. The mean RTD spread, o^2e, values changed significantly when the water level changed from low to high. However, the change for the RTD centroid of the first moment, λ_t, shows not much of a change in the values.

This study concluded that the residence time distribution changed significantly with the change of water levels but not so much with different flow rates. Moreover, a little change in the volume can influence the RTD characteristic in a constructed wetland because the volume is a function of depth. Depth or volume - of water effects the hydraulic efficiency of a constructed wetland and should be considered during the designing of wetlands [36]. Poor designing or sizing of wetland can affect its treatment performance. From this study, it can be stated that with the change of depth and volume in wetlands, the treatment process will be affected therefore the water quality can be assumed to be poor.

	Flow rates		Water levels	
Parameters	**High flows**	**Low flows**	**High flows**	**Low flows**
Peak concentration time. λ_p	0.25	0.22	0.19	0.29
Min. travel time of tracer dye, t_m	0.12	0.11	0.15	0.077
RTD centroid of first moment, λ_t	0.53	0.51	0.49	0.55
Normalized variance of RTD, σ^2_θ	0.65	0.55	0.73	0.47
All values are unitless.				

Table 1.
Summarized results for flow rates and water level (using high and low flows) [36].

6. Modeling of existing wetlands

6.1 Accuracy of modeling

The modeling of stormwater pollutants in runoff and the modeling of Stormwater Quality Treatment (SWQT) assets has developed significantly over the years. The first modeling primarily consisted of relatively simple mathematical equations which were considered somewhat crude, now software programs model many of the complex interactions which occur through the stormwater runoff and treatment process, via user-friendly interfaces [37]. Within Australia, the most widely used SWQT modeling program is MUSIC [38]. MUSIC is a stochastic model which utilizes probability to help determine the pollutants in stormwater runoff and the performance of SWQT assets [39]. Consistent with the rest of Australia, in the Melbourne region MUSIC is also the SWQT modeling software of choice. To approve new SWQT assets and to assess whether a new development is meeting best practices, Melbourne Water and Councils throughout Melbourne require a MUSIC model [40].

However, even though MUSIC models are a requirement by Councils and the major water authority within the Melbourne region, some research suggests that MUSIC models are not completely accurate and may over-treat or under-treat depend on the situation [38, 41]. In their paper Modeling stormwater treatment systems using MUSIC: Accuracy [41] undertook a series of comparisons between existing SWQT assets (located in Australia, Sweden, and New Zealand} and MUSIC models which had been created, with modified parameters to represent the actual conditions (e.g., inflows concentrations) and existing assets. They found that depending on the type of treatment asset, the accuracy of the MUSIC model may vary; in some cases, the MUSIC models overestimated treatment whereas in other cases the modeling under-estimated the treatment. It should be noted that the study did not investigate MUSIC's ability to accurately model wetlands, a topic which requires further examination, however, it does call into question MUSIC's ability to create accurate models and may provide a portion of the answer to the research question.

The uncertainty of the accuracy of MUSIC SWQT modeling could be due to setting up and calibrating the model. Several studies have found that some of the parameters in MUSIC, e.g., soil storage and field capacity, are crucial for obtaining accurate results and require calibrating based on local data [38, 39]. When creating MUSIC models, errors may occur in the modeling when the user uses the default MUSIC parameters, accidently inputs the incorrect parameters, is unaware of the correct parameters to input, or intentionally inputs the wrong values. To mitigate these potential human errors and improve the quality of MUSIC modeling, Melbourne Water produced MUSIC guidelines, which state the parameters to be used when creating a MUSIC model and general information about SWQT modeling elements. Whilst Melbourne Water's MUSIC Guidelines provide some recommended parameters, [38, 42] recommend further research into assessing the parameters required in MUSIC for catchments with "similar land use, climatic characteristics and hydrological behavior".

6.2 Condition of the wetland

The condition of the wetland could contribute to why modeling results may differ from wetlands. Wetland conditions that may influence the performance of the wetland include bad construction, outlet blockages, modification of the terrain by animals, etc. [6, 37]. Models must make some assumptions and they generally assume that assets will function as in intended. However, this is not always the case as there are many operational factors that affect the functioning of a wetland.

6.3 Timing of water sampling

To assess whether a comparison between the modeling and real-life conditions should occur, the first aspect that has to be determined is if the wetland is fully developed. Kadlec and Wallace [37] state that it can take more than 2 years for the wetland to develop fully. This is the duration required for the bio-system to mature, which requires amongst many other things the build-up of a layer of plant detritus over the base of the wetland so that congregations of periphyton and bacteria can form. The congregations of periphyton and bacteria are essential as these organisms' form part of the nutrient removal process [37]. Thus, if sampling is undertaken before the wetland is fully developed, it may not be reflective of the future potential of the wetland as the bio-system has not matured and is not working to full capacity.

Another aspect that is of importance to the timing of the sampling, is the change of seasons. As plants are seasonal and sprout and perish on an annual basis, the natural biological process dictates that there will be fluctuations in concentrations of nutrients due to the cycling, uptake and release of nutrients, by the plants [37]. Spring generally produces higher uptake of nutrients as the plants are growing and absorb more nutrients in this period whereas in autumn plants are generally dyeing-off and their decaying litter releases nutrients in the waterways [37]. As a result, depending on the timing of the sampling, the same fully developed wetland may produce significant results. However, [12] found that in their two-year study covering all seasons, there was no increase in nitrogen concentrations over the autumn and winter period, which raised the question that there may be other nitrogen removing mechanisms at play. Although, to mitigate this potential error, it is suggested that long-term sampling occurs so that a baseline performance can be determined which takes into account seasonal fluctuations [43].

6.4 Maintenance

One aspect which plays a crucial role in the condition of the wetland is maintenance. Regular maintenance is vital to the performance of a wetland as it facilitates the correct functioning of the wetland [37]. One maintenance task which enables the proper functioning of a wetland, is the cleaning-out of the sediment pond [16]. If the sediment pond fills up beyond its designed depth, the sediment pond may not have the required depth for sediment to settle and therefore the sediment will remain suspended in the water [37]. This suspended sediment may flow into the macrophyte zone and settle, or it may remain suspended and resist treatment. This has various implications to the effectiveness of the wetland, one being that if sediment settles on the base of the macrophyte zone and builds up, the water velocities in the macrophyte zone may increase due to the smaller flow area, which may cause erosion or impact on detention times [37]. Additionally, the makeup of plants in the macrophyte zone may change due to the lack of habitat, e.g., there may be no deep marsh plants, due to sediment changing the makeup of this zone to the shallow marsh.

Another way in which maintenance can impact the performance of a wetland is through the outlet structure. The outlet structure may become blocked due to litter or the natural decay of plant species. This blockage may result in the water level to rise for extended periods of time which can kill off a number of the plant species that inhabit the wetland [37]. Similarly, without regular maintenance of the wetlands, certain plants such as Typha may grow rampant and effect the hydraulic efficiency of the wetland. In doing so it may cause the water level to rise and kill off several plants [37]. To prevent this, it is suggested that scheduled maintenance of the wetland is to occur.

6.5 Frequency of water quality sampling and testing

Another area that may add to the differences between modeling and existing wetlands is in the sampling and testing. There are many aspects that need to be controlled to obtain accurate results. An important factor which must be taken into consideration to obtain an accurate result is frequency of the sampling. In one study it was found that "to sample TSS adequately within a storm event, at least 12 flow-weighted samples were required, and that polluto graphs of seven storm events needed to be sampled within a year to estimate mean annual loads at a reasonable level of accuracy" [43]. Building on this research [43] recommend that to have an error of less than 10% for sediment sampling, sampling must occur every three days or less for TSS–TN and TP will be different.

6.6 Samples' contamination

Differences between the modeling and actual results may result from the contamination of samples. This could occur in a multitude of ways hence the samplers must be vigilant and follow the protocols when proceeding to take samples. The following are some examples of ways that contamination may occur [44]:

1. Through disturbing the sediment/wetland base when sampling by placing the sample container too deep.

2. Dirty hands or instruments that take the sample may contaminate the water when sampling;

3. Sampling equipment is contaminated either through poor cleaning or coming into contact with other media before sampling.

7. How does government policy influence the sizing of wetlands?

Whilst there are no formal policies requiring wetlands there are policies that promote stormwater quality treatment to meet mandatory sediment and nutrient concentrations defined in statutory policies. The Environmental Protection Agency (EPA) has created the State Environment Protection Policies (SEPP), and of relevance to wetland policy is the SEPP Waters of Victoria [8]. The SEPP Waters of Victoria outline the required concentrations of sediment and nutrients in waterways and larger receiving bodies, such as Port Phillip Bay and Western Port, for waters to be considered healthy. In addition to the main policy, there are various Schedules that provide unique requirements for specific catchments, e.g., SEPP Schedule F6 relates to the water quality requirements of waters for Port Phillip Bay, SEPP Schedule F7 relates specific requirements of the Yarra catchment, etc. These policies are statutory under Section 16 of the Environment Protection Act 1970 [8]. The information provided in the SEPPs indicates what the receiving water's concentrations of pollutants should be [45]. Whilst this information is useful for providing guidelines for testing it provides little guidance on what stormwater quality treatment is required for new urban developments to maintain these concentrations in the receiving bodies [45]. To try and bridge this gap, the EPA and a panel of stakeholders and experts, which included the Department of Sustainability and Environment, Melbourne Water, Municipal Association of Victoria and local government, were engaged to develop the Best Practice Environmental Management (BPEM) guidelines, which provides a pragmatic methodology for maintaining the concentrations of sediment and nutrients listed in the SEPPs [8, 9].

Rather than produce guidelines that have a focus on concentrations, the BPEM guidelines promote performance objectives that utilize a sediment and nutrient load reduction procedure, and if followed, should maintain the concentrations listed in the SEPPs [9, 45]. **Table 2** defines the required load reductions by SWQT assets to meet the BPEM objectives and subsequently meet the SEPP (Waters of Victoria) concentration requirements.

The reduction loads were determined by the Cooperative Research Centre for Catchment Hydrology, in their research publication Best Practice Environmental Management Guidelines for Urban Stormwater. This research publication was based on data analysis within the Background Report to the Environment Protection Authority, Melbourne Water Corporation and the Department of Natural Resources and Environment, Victoria [10]. The primary intention was to create performance objectives that helped achieved the SEPP however, [10] believed that the BPEM performance objectives should be:

- Simple to use,
- Practical and cost-effective
- Prescriptive
- encouraging innovation
- flexible and
- justifiable and defensible (based on sound scientific method)
- "equitable and applicable to all organisations or communities who discharge to urban stormwater"

Pollutant	Receiving water objective	Current best practice performance objective
Post-construction phase		
Suspended solids (SS)	comply with SEPP (e.g., not exceed the 90th percentile of 80 mg/L)[1]	80% retention of the typical urban annual load
Total phosphorus (TP)	comply with SEPP (e.g., base flow concentration not to exceed 0.08 mg/L)[2]	45% retention of the typical urban annual load
Total nitrogen (TN)	comply with SEPP (e.g., bate flow concentration not to exceed 0.9 mg/L)[2]	45% retention of the typical urban annual load
Litter	comply with SEPP (e g No litter in waterways[1]	70% reduction of typical urban annual load[3]
Flows	Maintain flows at pre-urbanization levels	Maintain discharge for the 1.5-year ARI at pre-development levels
Construction phase		
Suspended solids	comply with SEPP	Effective treatment of 90% of daily run-off events (e.g., <4 months ARI). Effective treatment equates to a 50%ile SS concentration of 50 mg/L
Litter	comply with SEPP (e.g., No litter in waterways)[1]	Prevent litter from entering the stormwater system.
Other pollutants	comply with SEPP	Limit the application, generation and migration of toxic substances to the maximum extent practicable

[1]*An example using SEPP (Waters of Victoria 1988), general surface waters segment.*
[2]*SEPP Schedule F7—Yarra Catchment—urban waterways for the Yarra River main stream.*
[3]*Litter is defined as anthropogenic material larger than five millimeters.*

Table 2.
BPEM reduction load targets [9].

With these considerations in mind, one of the influencing factors on the performance objectives was land-take. Mudgway [10] found that an asset footprint of approximately 1% of the catchment was sufficient to produce reasonable reductions, e.g., TSS (40–80%) and T.P (35–45%). Hence the performance objectives were created in an attempt to satisfy the SEPP and the ideals above and not necessarily to prescribe treatment for certain flow frequencies. SEPP and BPM have an influence on the sizing of SWQT assets however it is indirectly and not through explicit statements. Most wetlands are sized for the 1 in 3 months flow, however, this appears to be more of a rule of thumb, which can be enforced by local authorities such as Melbourne Water, rather than a statutory requirement [17]. The 1 in 3-month flow is nominally sized flow frequency which has generally been utilized to satisfy the BPEM performance objectives (retention of the typical annual load: TSS = 80%, TP = 40%, TN = 45%) [9]. The primary objective of wetland sizing is to meet the BPEM performance objectives, and it just SQ happens that the 1 in 3-month flow meets this objective.

Despite the fact the BPEM performance targets are the primary drivers for sizing wetlands, local authorities such as Melbourne Water may enforce that wetland be sized for 1 in 3-month flows. This may not necessarily be due to treatment meeting the SEPP concentrations but rather for maintenance reasons, e.g., plant protection [17]. The authority with which Melbourne Water may influence the sizing of the wetland is somewhat convoluted as they receive their authority from the Water Act 1989, and the Environment Protection Act 1988 via the SEPP (Waters of Victoria). Through these acts and policies, Melbourne Water has the authority to dictate the design parameters of wetlands and in their most recent wetlands manual they state "All flows ≤ the peak three-month ARI event is transferred into the macrophyte zone when the EDD in the macrophyte zone is at Natural Water Level (NWL) (**Figure 2**)" [17].

8. Conclusions and recommendations

Based on the review of literature it can be concluded that:

1. Constructed wetlands are cost-effective treatment systems that can be used to treat urban stormwater runoff.

2. Wetlands plants that are frequently flooded can have an adverse effect on the treatment process.

3. The change of depth and volume of water in constructed wetlands affects the treatment process.

4. It can take more than 2 years for the constructed wetland to fully develop, when the bio-system matures, and form part of the nutrient removal process. Thus, sampling and monitoring before the wetland is fully developed should be taken into consideration to reflect the future potential of the wetland.

5. To have an error of less than 10% for sediment sampling, sampling must occur every three days or less for TSS.

6. Maintenance is important for the ongoing effectiveness of constructed wetlands in water treatment and in flood mitigation.

7. Whilst there are no formal policies requiring constructed wetlands, in Victoria, Australia they are typically installed as part of large-scale urban developments to comply with the BPEM guidelines.

8. Further research into assessing the parameters required in MUSIC for catchments with "similar land use, climatic characteristics and hydrological behavior" is recommended.

Author details

Shirley Gato-Trinidad[1*], Jamie Carroll[2], Ezreena Aladin[3] and Tristan Gilbert[4]

1 Swinburne University of Technology, Hawthorn, Australia,

2 SPIIRE (Integrated Water), Australia

3 DCE, Australia

4 Integral Group Australia, Australia

*Address all correspondence to: sgatotrinidad@swin.edu.au

References

[1] Malaviya P, Singh A. Constructed wetlands for management of urban stormwater runoff. Critical Reviews in Environmental Science and Technology. 2012;**42**(20):2153-2214

[2] Sauter G, Leonard K. Wetland design methods for residential wastewater treatment. JAWRA Journal of the American Water Resources Association. 1997;**33**(1):155-162

[3] Brix H. Wastewater treatment in constructed wetlands: System design, removal processes, and treatment performance. In: MG A, editor. Constructed Wetlands for Water Quality Improvement. Florida: CRC Press; 1994. pp. 9-22

[4] Carleton JN, Grizzard TJ, Godrej AN, Post HE. Factors affecting the performance of stormwater treatment wetlands. Water Research. 2001;**35**(6):1552-1562

[5] Acreman M, Holden J. How wetlands affect floods. Wetlands. 2013;**33**(5):773-786

[6] Hunt W, Greenway M, Moore TC, Brown R, Kennedy SG, Line D, et al. Constructed storm Water wetland installation and maintenance: Are we getting it right? Journal of Irrigation and Drainage Engineering-ASCE. 2011;**137**(8):469-474

[7] Lucas R, Earl ER, Babatunde AO, Bockelmann-Evans BN. Constructed wetlands for stormwater management in the UK: a concise review. Civil Engineering and Environmental Systems. 2014;**32**(3):1-18

[8] EPA, EPAV. State Environment Protection Policies. Victoria, Australia: EPA; 2012. Viewed 21/05/2017. Available from: http://www.epa.vic.gov.au/about-us/legislation/state-environment-protection-policies

[9] CSIRO. Urban Stormwater: Best Practice Environmental Management Guidelines. Melbourne: CSIRO; 1999

[10] Mudgway LB, Duncan HP, McMahon TA, Chiew FHS. Best Practice Environmental Guidelines for Urban Stormwater, Background Report. Cooperative Research Centre for Catchment Hydrology. 1997

[11] MangangkaIR,EgodawattaP,ParkerN, Gardner T, Goonetilleke A. Performance characterization of a constructed wetland. Water Science and Technology. 2013;**68**(10):2195-2201

[12] Wadzuk BM, Rea M, Woodruff G, Flynn K, Traver RG. Water-quality performance of a constructed stormwater wetland for all flow conditions. JAWRA Journal of the American Water Resources Association. 2010;**46**(2):385-394

[13] Walsh O, Fletcher TD, Ladson AR. Stream restoration in urban catchments through redesigning stormwater systems: Looking to the catchment to save the stream. Journal of the North American Benthological Society. 2005;**24**(3):690-705

[14] Semadeni-Davies A, Hernebring C, Svensson G, Gustafsson L-G. The impacts of climate change and urbanization on drainage in Helsingborg, Sweden: Suburban stormwater. Journal of Hydrology. 2008;**350**(1):114-125

[15] Melbourne Water, C. Drainage Schemes Explained, Viewed 27/05/2017, 2016. Available from: https://www.

melbournewater.com.au/Planning-and-building/schemes/map/Pages/Drainage-schemes.aspx

[16] Allison R, Francey M, Csiro & Melbourne Water, C. WSUD Engineering Procedures: Stormwater. Collingwood, Victoria, Australia: CSIRO: Melbourne Water; 2005

[17] Melbourne Water, C. Constructed Wetlands Design Manual - Part A2 - Deemed to Comply Design Criteria. Melbourne: Melbourne Water Corporation; 2015

[18] Brix H. Do macrophytes play a role in constructed treatment wetlands? Water Science and Technology. 1997;**35**(5):11-17

[19] Gersberg RM, Elkins BV, Goldman CR. Nitrogen removal in artificial wetlands. Water Research. 1983;**17**(9):1009-1014

[20] Reddy K, Patrick W, Lindau CW. Nitrification-denitrification at the plant root-sediment Interface in wetlands. Limnology and Oceanography. 1989;**34**(6):1004-1013

[21] Vaze J, Chiew FHS. Nutrient loads associated with different sediment sizes in urban stormwater and surface pollutants. Journal of Environmental Engineering. 2004;**130**(4):391

[22] Ibekwe AM, Lyon SR, Leddy M, Jacobson-Meyers M. Impact of plant density and microbial composition on water quality from a free water surface constructed wetland. Journal of Applied Microbiology. 2007;**102**(4):921-936

[23] Fitch MW. Constructed Wetlands. Amsterdam, Netherlands: Elsevier; 2014. pp. 268-295

[24] Hey D, Philippi N. Flood reduction through wetland restoration: The upper Mississippi River basin as a case history. Restoration Ecology. 1995;**3**:4-17

[25] Galloway GE. Learning from the Mississippi flood of 1993: Impacts, management issues, and areas for research. In: U.S. Italy Research Workshop on the Hydrometeorology, Impacts, and Management of Extreme Floods. United States: National Defense University (USA).; 1995

[26] Potter K. Estimating potential reduction flood benefits of restored wetlands. Water Resources. 1994;**97**:34-38

[27] Galat DL, Frederickson LH, Humburg DD, Bataille KJ, Bodie JR, Dohrenwend J, et al. Flooding to restore connectivity of regulated, large-river wetlands. (Lower Missouri River) (Flooding: Natural and Managed Disturbances). BioScience. 1998;**48**(9):721

[28] Greenway M. Constructed wetlands for water pollution control and ecological health. A challenge for environmental engineers. International Symposium on Ecosystem Health. Brisbane, Australia: Transdisciplinary Approaches to Ecosystem Health; 2000

[29] Kadlec RH. Overland flow in wetlands: Vegetation resistance. Journal of Hydraulic Engineering. 1990;**116**(5):691-706

[30] Jadhav RS, Buchberger SG. Effects of vegetation on flow through free water surface wetlands. Ecological Engineering. 1995;**5**(4):481-496

[31] Ogawa H, Male JW. Simulating the flood mitigation role of wetlands. Journal of Water Resources Planning and Management. 1986;**112**(1):114-128

[32] Shultz S, Leitch J. The Feasibility of Wetland Restoration to Reduce Flooding in the Red River Valley; A Case Study of

the Maple River Watershed, Agribusiness and Applied Economics Report No. 432a. North Dakota, USA. 2001

[33] Casanova M, Brock M. How do depth, duration and frequency of flooding influence the establishment of wetland plant communities? Plant Ecology. 2000;**147**(2):237-250

[34] Brock T, Velde G, Steeg HM. The effects of extreme water level fluctuations on the wetland vegetation. Archiv für Hydrobiologie -Beih. Ergebn. Limnol. 1987;**27**:57-73

[35] Werner TM, Kadlec RH. Wetland residence time distribution modeling. Ecological Engineering. 2000;**15**(1):77-90

[36] Holland JF, Martin JF, Granata T, Bouchard V, Quigley M, Brown L. Effects of wetland depth and flow rate on residence time distribution characteristics. Ecological Engineering. 2004;**23**(3):189-203

[37] Kadlec RH, Wallace S. Treatment Wetlands. 2nd ed. Hoboken: Elsevier; 2008

[38] Dotto CBS, Deletic A, McCarthy DT, Fletcher TD. Calibration and sensitivity analysis of urban drainage models: MUSIC rainfall/runoff module and a simple stormwater quality model. (Technical paper). Australian Journal of Water Resources. 2011;**15**(1):85

[39] Ellis JB, Viavattene C. Stormwater modelling and sustainable management in highly urbanized areas. In: Eslamian S, editor. Handbook of Engineering Hydrology Environmental Hydrology and Water Management. Hoboken: CRC Press; 2014

[40] Melbourne Water Corporation. STORM and MUSIC Tools. Australia: Melbourne Water Corporation; 2016. Viewed 27/05/2017. Available from: https://www.melbournewater.com.au/Planning-andbuilding/Stormwater-management/tools/Pages/default.aspx

[41] Imteaz MA, Ahsan A, Rahman A, Mekanik F. Modelling stormwater treatment systems using MUSIC: Accuracy. Resources, Conservation & Recycling. 2013;**71**:15-21

[42] Melbourne Water Corporation. MUSIC Guidelines. Australia: Melbourne Water Corporation; 2016

[43] Fletcher T, Deletic A. Statistical observations of a stormwater monitoring programme; lessons for the estimation of pollutant loads. In: Proceedings of NOVATECH 2007, 6th International Conference on Sustainable Techniques and Strategies in Urban Water Management. 15-28 June 2007, Lyon, Rhone-Alpes, France. 2007

[44] NWQMS, DoEaE. National Water Quality Management Strategy: Australian Guidelines for Water Quality Monitoring and Reporting. Canberra, A.C.T: Australian and New Zealand Environment and Conservation Council; 2000

[45] EPA, EPAV. Consolidated SEPP Waters of Victoria. EPA: Victoria; 1988

Chapter 2

Flood Risk Assessment in Urban Areas: The Historic City Centre of Aveiro as a Case Study

Mayra Alejandra Estrella Núñez, Alkmini Firtinidou-Stergiou, Margherita Rago, Chenxin Jonathan Yee, Alberto Barontini, Tiago Miguel Ferreira and Daniel V. Oliveira

Abstract

Floods are among the most frequent and widespread natural hazards worldwide, with historic buildings proving to be particularly vulnerable. This work focuses on the large-scale flood risk assessment of the Historic City Center of Aveiro in Portugal. Due to the system of canals passing through the center and the extreme proximity to a lagoon, Aveiro is extremely prone to flooding. Furthermore, considering the great historical and artistic value the city center retains, its selection as a case study for flood risk assessment is fully supported. The work implements a recently developed methodology, combining hazard and physical vulnerability indicators to classify risk and define intervention priorities. Subsequent to an extensive survey and evaluation of almost 500 buildings, the raw data collected was classified using the Geographic Information System (GIS) tool. Through the macro-scale risk assessment, an essential insight is provided into the way each building is affected by flood events and, based on this knowledge, strategic rehabilitation interventions can be prioritized. To validate results, the work proposes a comparison with an analogous case study, and finally reflects upon the effective risk management and mitigation proposals as well as possible adaptations of the methodology for future applications.

Keywords: flood risk assessment, macro-scale risk assessment, flood vulnerability, geographic information system, historic center, heritage constructions

1. Introduction

Floods are among the most widespread climate-related [1] and frequently occurring natural hazards. In the past decade alone, the annual reported economic losses from floods have reached tens of billions of US dollars, with their repercussions affecting thousands of people. More specifically, historic city centers and heritage buildings are particularly vulnerable to the destruction caused by flooding. This can be attributed to the fact that ancient settlements tend to be located within the vicinities of water sources, attesting to the important role water plays in the development of these communities. Unfortunately, this also means that the locations of present-day heritage structures and buildings are primarily within flood-prone areas, giving them a preexisting disadvantage of being exposed to hazardous conditions.

All these reasons demonstrate the necessity to develop further studies and methods to understand and manage the flood risk, especially for historic sites and structures. They have the propensity to present certain singularities, such as construction typologies and traditional building materials, that can lead to an increase in their vulnerability to flooding hazards. The potential loss of buildings with high and irreplaceable cultural value, as well as the consequent economic impacts from which a city could suffer, stresses the urgency to develop systems of indicators for managing the risk to heritage assets. These systems aim to provide valuable information for the assessment of the impact of natural hazards on cultural, social, economic, and environmental conditions. As such, historic centers are critical areas of study and deserve special attention.

To this end, the present study discusses the application of a recently developed flood risk-assessment approach that combines flood hazard and building vulnerability indicators to identify and classify risk while also defining intervention priorities. This methodology [1] is implemented in the Historic City Center of Aveiro, Portugal, particularly in the Beira Mar neighborhood, one of the most picturesque of the city. Due to its proximity to the Ria de Aveiro and the system of canals throughout the historic center, the city is highly prone to flooding. The selection of Aveiro's historic center as the subject of this large-scale risk assessment is further supported by the city's unique history and architecture, widely recognized to be of high heritage value and significance.

This study involves the survey and evaluation of 495 buildings, completed on-site as well as through remote survey methods, which are usually the go-to methods for the vulnerability assessment of post-earthquake events [2–5]. The results were integrated into a Geographic Information System (GIS) tool, alongside prepared flood hazard data, for subsequent analysis of the respective flood vulnerability and risk. Through this macro-scale risk assessment, it is hopeful that valuable insight is generated into the manner in which each building is affected by the flood events. Discussions also include proposals of effective risk management and mitigation strategies, comparison of results with an existing case study, as well as adaptions of the methodology for utilization in different contexts.

2. Methodology

The analysis presented herein has been carried out using the methodology proposed by Miranda and Ferreira [1]. This particular methodology is adjusted so as to be implemented as a micro-scale assessment. In this case, the elements of interest are the historical buildings and the losses, which are mainly defined in terms of physical damage caused to the building fabric, namely, the direct tangible damage. For historic buildings, this damage is deemed the most severe. Moreover, to deal with historic sites, the method addresses direct intangible damage, in terms of losses to the cultural value of heritage buildings.

The vulnerability in this methodology refers to the intrinsic characteristics of the building. Other factors define the social, historical, and economic value of the asset and its interaction with the water. The results can be seen as a synthetic vulnerability function as they are not derived from damage data after flood events. The vulnerability index is a relative measure of the expected severity of the physical

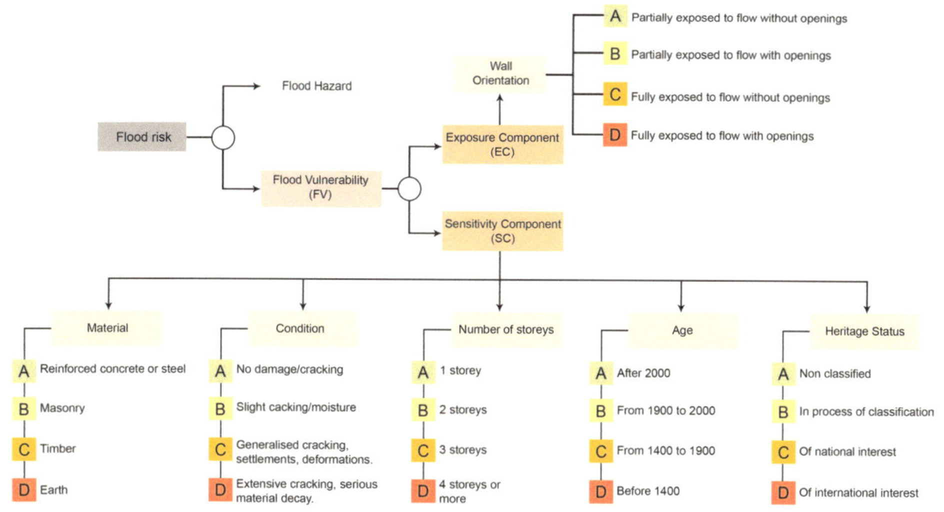

Figure 1.
Diagram of methodology [1].

impact of the flood on the building in comparison to other assets ranked. The value of the vulnerability index does not account for the hazard and only partially accounts for the exposure, following an approach that is becoming common in flood risk assessment [6, 7].

As shown in **Figure 1**, the methodology defines flood vulnerability (FV) as the product of the exposure component (EC) and the sensitivity component (SC). Five parameters are considered to define the sensitivity component, each one of them containing four attributes. The exposure parameter refers to the exposure of the building to the flood, which encompasses aspects like the orientation of the most exposed façade to the water flow or the location and dimension of the openings—doors and windows [1].

2.1 Flood vulnerability

The goal of vulnerability assessment is to understand how a system will be affected by floods. Even though there are different classes of vulnerability, physical vulnerability is the most commonly assessed because it is the easiest way to characterize it. Furthermore, the nonphysical vulnerability is usually evaluated in terms of economic loss, which is related to the interruption of various activities and is often much more challenging to estimate.

According to Balica [8], flood vulnerability can be determined by a physical approach, where hydrological models are used to estimate flood hazard and economic consequences. Moreover, Balica also mentions an empirical approach related to quantitative or qualitative indicators to rate the vulnerability without inputting the hazard intensity. Regarding the methodology adopted for the present study, no attribute has been assigned the value 0, since it is considered that every historic-building feature measured indicates some degree of vulnerability to the hazard. However, for all the descriptors to contribute equally, a lower boundary of 10 is applied to the rating scale for them, except for the land slope [9].

Based on the estimated individual parameters, the flood vulnerability is computed according to Eq. (1).

$$\text{Flood Vulnerability (FV)} = \text{Exposure Component (EC)} \times \text{Sensitivity Component (SC)} \quad (1)$$

2.2 Exposure component

Exposure analysis aims to examine the economic assets and activities influenced by the flood. Exposure can be determined by geospatial mapping to identify the location of the assets of interest related to the flood hazard. In addition to this, the exposure contemplates some intrinsic characteristics of the elements analyzed.

2.2.1 Wall orientation

Wall orientation assesses the position of the asset with respect to the expected water flow. This parameter considers that the wall in a plane perpendicular to the water stream direction (fully exposed) presents a bigger pressure load than the ones that are simply immersed. It also considers the presence of openings reachable by the water. The characterization of the buildings as fully or partially exposed without openings is attributed to the ones where the windows and doors are not facing the flow of the water [10]. The assessment is based on the orientation and the characteristics of the main façade. In the case of buildings located in low-lying areas, the class is increased to C or to D for partially and fully exposed buildings.

2.3 Sensitivity component

2.3.1 Material

This parameter considers the lateral capacity of the structure, the resilience to ground movement, and the absorption capacity, factors that contribute to the risk of collapse in the short term and degradation in the longer term. In the case of degradation, it depends on two factors: the consequences of improper drying and on the contaminants and salts present in the flow [11].

Among materials, earth is considered the most vulnerable. The materials' porosity determines their absorption capacity, but it can also promote their resilience, as it allows them to tolerate heavy wetting. It should be mentioned that the unbaked earthen structures are the most likely to present disintegration caused by the wetting and water pressure [12].

Masonry buildings are less prone to suffer collapse. However, they can present moisture in the elements and different types of decay such as spalling or exfoliation induced by the presence of salts and contaminants. Previous experiences demonstrated reduction up to 50% of the capacity in dry condition. Therefore, masonry structures should be properly inspected post-event [10].

Modern materials such as steel and concrete are less vulnerable. Moreover, the reparation and replacement of traditional materials may be more expensive. Timber elements are prone to biological attacks, such as fungi and insects. After the flood event, if the element is properly dried, serious damage is not expected. Moreover, saturated elements undergo significant deformation.

2.3.2 Number of stories

It is argued that single-story buildings are the most vulnerable, since they do not possess a flood-free story to facilitate the evacuation [13]. Nonetheless, in terms of stability, higher buildings with the same footprint and superficial foundations are more susceptible to the effects of the flood on the subsoil, such as differential settlements [5]. Therefore, for the present study, the vulnerability is considered as increasing with increasing number of stories.

2.3.3 Damage and cracking

Damage can lead to higher water absorption and lower strength in the materials, such as in the case of wood. Indeed, water ingress can easily occur through cracks, and once inside the building, water can damage movable assets and valuable nonstructural elements, contributing significantly to the economic losses. Therefore, restoration and maintenance measures are essential to prevent water penetration. This indicator includes an assessment of the state and quality of the interventions that can be carried out [6].

2.3.4 Age

This parameter may represent more valuable assets. This is valid for countries where historic structures have a higher real-estate value, being thus inhabited by higher-income residents [14]. Furthermore, older buildings are likely more prone to suffer flood damage. It is worth noting that the four levels of the age indicator may require an adaptation to the local evolution over time of the investigated building stock.

2.3.5 Heritage status

The heritage status is directly related to the socioeconomic impact that the preservation of these assets could have. For this reason, a higher value implies higher exposure of the asset. Moreover, the buildings that are listed are often subjected to more requirements, which results in higher intervention costs. In these cases, it is also important to contemplate the nonstructural value of artworks and other movable assets that are hosted in the building, a fact that will increase their vulnerability [15].

3. Case study: the Historic City Centre of Aveiro

The city of Aveiro is located on the shore of the Atlantic Ocean, in the Central region of Portugal. The territory is administratively designated as Beira Litoral, with Aveiro being the capital of the district and the name of the municipality (**Figure 2a**). The city has a population of 80,880 and a total area of 197.58 km^2.

The city's location (**Figure 2b**) has unique characteristics: while to the east it is confined by mountains, to the north, south, and west spreads the immense lagoon to which the city gives its name—Ria de Aveiro—and that distinguishes the surrounding scenery stretching out to the sea. In an area geographically and historically isolated from the mainland communication routes, the estuary of the numerous waterways that flow into it has, over the centuries, been the determining factor in the movement

(a)

(b)

Figure 2.
Aveiro's city centre (a) and entrance of the Barra da Ria de Aveiro (b). Sources: Getty images/Google Earth.

of people and goods. In particular, the river Vouga, the largest of the watercourses, presents an important communication route. The existence of the Ria de Aveiro is essentially linked to the one of the city. Today, it is composed of countless islands, lagoons, and channels; however, throughout the centuries, the natural substrate has undergone profound changes, which has eventually led to the formation of the estuary.

3.1 History and development of Aveiro

Nowadays, Aveiro is one of the most populated cities in Central Portugal, grown around its thriving port. Nevertheless, over the centuries, the process of consolidation and growth of Aveiro has gone through different phases, each reflecting the living conditions in the region and its relationship with the lagoon and the sea.

Initially known as *Avarium*, documentation seems to place the origin of the city at least back to the 10th century. The first centuries of Aveiro's existence are essentially marked by a period of consolidation of the primitive village, developed around the main church of St. Miguel, which assumed the status of a city only in the 13th century. Always linked to economic activities, Aveiro invested in the salt production and in the naval commerce as its most valuable assets, which constituted the main economic activities that accompanied its growth until the end of the 14th century [16].

Following the centuries, the building of the walls around the urban center and the development of the salt industry, agriculture, and fishing established for Aveiro a period of prosperity. However, the sedimentation in the "Barra" of the Ria de Aveiro (the link between the lagoon and the ocean) progressively led to a period of crisis that struck Aveiro until the 18th century, when the closure of the Barra became permanent. The construction of Barra Nova in 1808, along with the development of transport, especially the passage of the railway line Lisbon-Oporto, marked the beginning of a new era for the city [16].

This revival also coincided with the beginning of the industrialization process. During this phase, the city's expansion took place mainly in the northern part of the canal, with the formation of the Beira-mar district. From the last half of the 20th century onward, the population of Aveiro continued to grow steadily, with the creation of the university being a contributing factor.

3.2 Building typologies and materials

The predominant building typology in Aveiro is the so-called gothic-mercantile that, commonly, features living and dining rooms facing the street and kitchens placed at the rear of the house. All the rooms are connected by a long corridor perpendicular to the street. The buildings tend to have a single structural span, and they are inserted in narrow and deep lots attached to one another [17]. The height can vary from one story to a maximum of four, but the majority possesses two or three stories.

The influence of Art Nouveau is evident in the decoration of a significant number of constructions since the beginning of the 20th century. In the city of Aveiro, Art Nouveau was partially adapted to local traditions, creating a peculiar combination characterized by outstanding decorated façades and conservative interiors often built with poor local materials.

Indeed, the dominant construction technique relies on natural materials. Masonry with lime mortar was used for various buildings. Until the early 20th century, the Eirol stone was largely used for towers and aqueducts, whereas the Ança stone was commonly used for foundations, basements, and decorative elements [18].

Nevertheless, most of the buildings observed are characterized by the use of adobe blocks in the construction of the external walls. It has been estimated that around 25% of the existing buildings of Aveiro are made of earthen blocks, of which 40% are located in the city center, including most of the Art-Nouveau-style buildings. This is an extraordinary and distinctive feature of the city but also one of its main problems when it comes to conservation, with adobe being very vulnerable to water and considerably less resistant than stone. Only around the 1950s was adobe progressively replaced by reinforced concrete [19].

Along with adobe, the presence of pinewood can be observed in abundance, mainly for roofs, floors, and window frames of the buildings. It was implemented mainly with the tabique technique, one of the most traditional building techniques in Portugal. Less frequent but still present is the usage of brick, especially for public buildings [19].

3.3 Flooding history

The estuary area of the Ria de Aveiro is spread over about 11,000 hectares, more than 6000 of which is permanently covered with water. The surrounding area has low altitude and orographic relief, lacking natural protections against flooding. **Figure 3** shows the chronological distribution of the main urban-flood events occurred in the city of Aveiro.

Concerning the events before the 20th century, the documentation is scarce; therefore, fewer urban floods are recorded. However, it is known that a particularly great event occurred in 1739. Another one, in 1774, had terrible consequences as the Barra gave no outlet to the dammed waters and caused an epidemic that decimated the population.

In the 20th century, several flood episodes occurred. The most notable in terms of losses affected mainly the lower part of the city, with the historical city center being completely covered by water in 1937. Finally, severe floods struck Aveiro in 1955, 1957, 1964, and 1966, following strong storms and rise of water levels, and, more recently, in 1993, when the entire lower part of the city was inundated [20].

Figure 3.
Timeline of the major floods in Aveiro during the 20th century [20].

DOI: http://dx.doi.org/10.5772/intechopen.109867

4. Site inspection and fieldwork

The present study focuses on the Bairro da Beira-mar, one of the most central and picturesque neighborhoods in the Historic City Center of Aveiro, characterized by the presence of its distinctive adobe Art Nouveau buildings.

4.1 Survey area

The investigated area is approximately 118.000 m^2 and consists of 495 buildings, mostly low-rise, which are divided into regular and some more irregular blocks forming a grid system. The buildings are commonly characterized by attached, deep lots with a very narrow front (mostly less than 7.5 meters) and mainly two or three stories. Various significant buildings, including the listed heritage, are located in the area.

4.2 Site inspection and fieldwork

The study area was divided, for ease of survey, into five specific zones (**Figure 4**) according to the main streets. The survey of the zones was completed in two phases. Phase 1 comprises the on-site surveys of Zone A, B, and C (242 buildings). Phase 2 consists of remote surveys for Zone D and E (253 buildings), complemented with on-site validation of the buildings not accessible remotely via Google Earth and Street View, as provided by Google and Maxar Technologies.

The building survey was conducted using a checklist hosted on Google Forms. The building checklist includes the inputs for all the building parameters required (e.g., number of stories, construction material, heritage status, etc.) in order to fulfill the

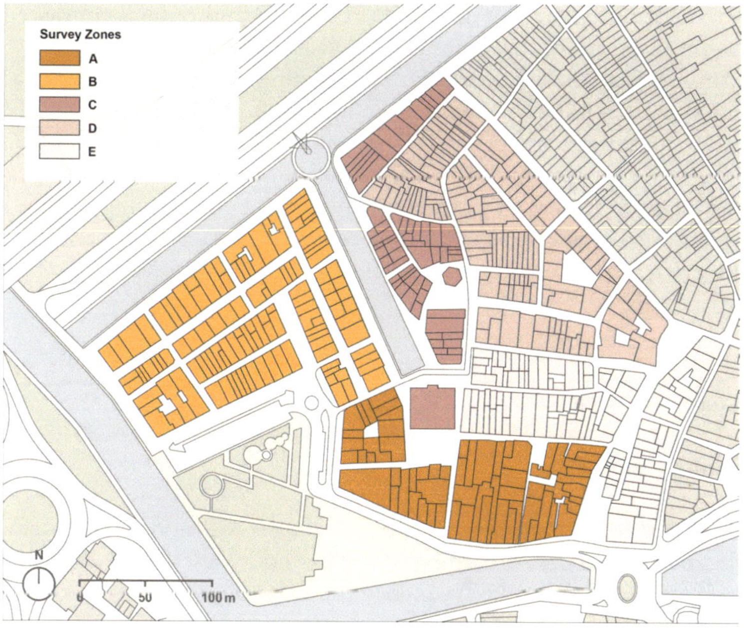

Figure 4.
Site survey zones.

(a) (b)

Figure 5.
Thermographic survey showing (a) reinforced concrete structure and (b) masonry units with 16 cm ruler as indicated by an asterisk.

methodology, as well as additional information (such as number of occupants, function of building, etc.) that could be helpful for further analyses.

The first on-site inspection was carried out on 14 and 15 November 2021 by a team of four persons, working in pairs. The team completed the survey of Zone A, B, and C during the first on-site visit, covering 242 buildings.

The remaining Zone D and E were surveyed remotely using Google Earth and Street View. Having visited the site in Phase 1 prior to conducting the remote survey, the team was well-acquainted and familiar with the urban morphology and building features to be able to replicate an equally objective judgment when conducting the survey remotely.

Due to the limitations of Google Street View, several streets were visually inaccessible. Thus, no information was available for the inspection of a handful of buildings, amounting to approximately less than 10% of Zone D and Zone E combined. These buildings were surveyed in a separate on-site visit on 15 March 2022.

During this on-site survey, thermal imaging using an FLIR T540 thermal camera was conducted to ascertain the construction materials of the selected buildings. In some cases, the thermography provided inconclusive results, depending on the façade coverings. Indeed, thick plastering and finishing with different materials and colors hindered the interpretation of the structural elements beneath. Nonetheless, in several instances, reinforced concrete structural members (**Figure 5a**) as well as masonry units (**Figure 5b**), which would not have been visible to the naked eye otherwise, were clearly identified.

5. Flood risk assessment

5.1 Vulnerability module

The on-site and remote surveys allowed for the classification of the parameters affecting flood vulnerability, according to the methodology described in Section 2. **Figures 6a, b** and **7a, b** show the frequency distribution and the spatial distribution concerning, respectively, two of the most relevant parameters that compose the flood vulnerability index: the wall orientation, which determines the exposure component of the vulnerability, and the condition of the building, which falls under the sensitivity

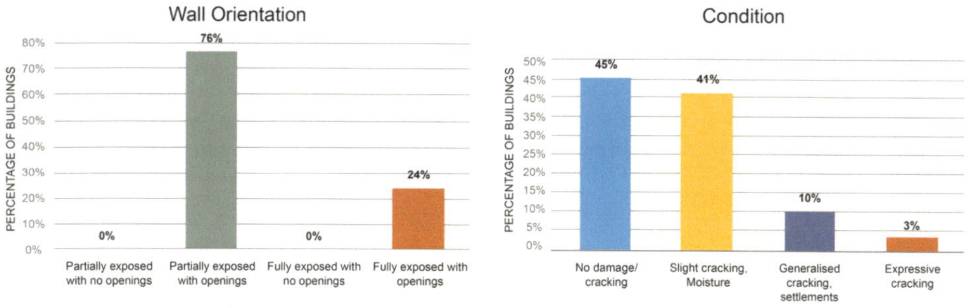

(a) (b)

Figure 6.
Frequency distribution of (a) wall orientation and (b) building condition parameters.

(a) (b)

Figure 7.
Spatial distribution of (a) wall orientation and (b) building condition parameters.

component. Regarding the former, all the investigated buildings fall into two categories, namely, "partial exposure with openings" or "full exposure with openings", with most of the buildings that present high level of exposure (class D) being located closer to the canal.

For what concerns the buildings' condition, their conservation state can be deemed as satisfactory. Almost 86% of the buildings assessed either are in a good condition (about 45%) or present minor conservation issues, mainly related with small cracks and moisture issues. The remaining 13% of structures face more serious degradation such as significant cracking, material loss, or settlement.

Regarding the heritage status, even though a great amount of the present building complexes are of architectural interest, only 3% of them are listed heritage with local/municipal interest. To consider the intangible value of such assets, the methodology was modified, without altering its underpinning rationale, to include within category B the status of significant local heritage that was not included in the original formulation.

However, uncertainties arose for several buildings in the assessment of the remaining parameters, namely, construction material and age. This is due to the recent finishing that does not expose the original construction materials beneath and a lack of documentary evidence on the evolution of the building stock in the neighborhood. To address these uncertainties, several assumptions were made. Most of the buildings with uncertain age were assumed to be built between 15th and 19th centuries, that is, category C. This option not only corresponds to the largest time range and, consequently, the highest probability for a building to be built during it but also is in accordance with the period of evolution of the Beira Mar neighborhood, completed during the 20th century (category B), which would be a less conservative assumption. On the other hand, buildings dating back to before the 15th century (category D) as well as to the 21st century (category A) were easily identifiable, due to their morphological and structural characteristics. Based on the aforementioned assumption, 81% of the buildings were classified as being built between the fifteenth and twentieth centuries.

Regarding the material, masonry and earthen constructions were often hardly distinguishable. Therefore, two cases were conceived. Case 1 has all uncertain construction materials classified as B (Masonry), while Case 2 has all uncertain construction materials graded as D (Earth). For the flood vulnerability assessment, Case 1 is selected, as it is more realistic based on the results obtained from the on-site survey.

Upon the classification, a clear predominance of masonry buildings emerged, accounting for 67% of the assets, assuming them as made of stone or bricks in the first scenario and of adobe in the second one. Nineteen percent of the buildings made of reinforced concrete are newer constructions.

Following this, the geographical distribution of the index is presented in **Figure 8**. The highest FVI value obtained for both the scenarios is 64.6. While this value is not negligible, in both cases, a very small percentage of buildings present a medium or high vulnerability.

Comparing the assessment of two different scenarios, it can be observed that Case 2, as expected, results in a higher vulnerability. Nonetheless, the overall vulnerability is slightly sensitive to the uncertainty in the classification of the material, as demonstrated by the very similar results both in terms of FVI values and distribution. Therefore, only Case 1 is considered for the risk assessment hereafter. The vulnerability is rather affected by the exposure component, with the most concerning buildings located along the canals.

Finally, the buildings are divided into three categories, namely, low, moderate, and high vulnerability (**Figure 9**). Index values of 20 and 40 are set as thresholds [14].

(a) (b)

Figure 8.
Flood vulnerability index results: Distribution over the study area Case 1 (a) and Case 2 (b).

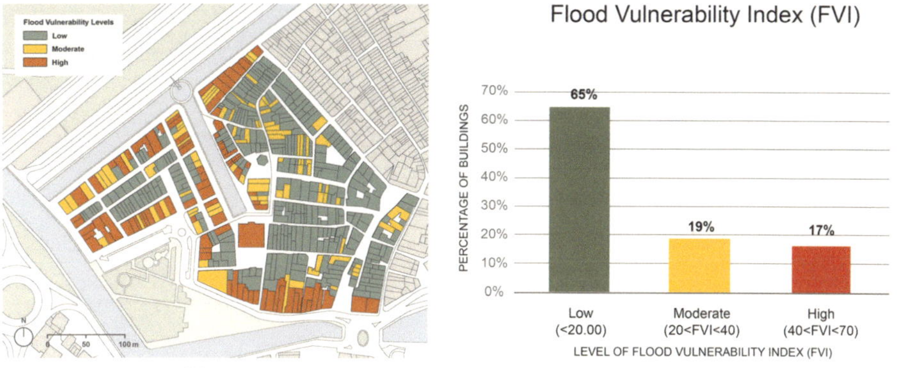

(a) (b)

Figure 9.
Distribution of flood vulnerability levels over the study area (a) and histogram with respect to the number of buildings (b).

5.2 Hazard levels

The hazard component of the flood risk assessment was analyzed in collaboration with the University of the West of England, which provided flood extents, velocities, and water depth maps associated with return periods of 20 and 100 years—that is, frequent and rare flooding events, as per the Portuguese Decree-law no. 115/2010 of 22 October. From these maps, the flood depth and velocity affecting each building were estimated, transforming the overlaid data (**Figure 10a**) into building-specific data (**Figure 10b** and **c**).

Within the investigated area, the estimated flood extent for 20- and 100-year return period is the same. Similarly, the flood velocity is equal in both cases, corresponding to 0.05 m/s. The flood depth, instead, significantly changes in the two scenarios. Indeed, for the 20-year return period, the buildings will be exposed to an average depth of 0.93 m and for the 100-year scenario, to an average depth of 0.98 m. Out of the 228 buildings affected by the flood in the two scenarios, a water height between 1.5 and 2.5 m is expected for 26 building under the 20-year return period and 28 under the 100-year return period. This is a significant value, but it represents a low percentage of buildings (5%) in comparison to the total amount.

Water height (y) and velocity (v) are used to calculate a single hazard indicator as follows:

$$H = y(v + 0.5) \tag{2}$$

Based on the classes of hazard, defined in the Portuguese Floods Directive, five hazard levels are considered, namely, negligible, low, moderate, high, and extreme as shown in **Table 1**.

According to the criterion adopted in this matrix-based analysis, the level of flood hazard throughout the study area does not alter much passing from the 20-year to the 100-year peak flow scenario. By observing the distribution of percentages, the results can be deemed as satisfactory, portraying a low level of danger. In terms of spatial distribution (**Figure 11a** and **b**), it is possible to identify two blocks that can be particularly affected: one in the northern part of the city center and one in the southern, always in the flood-prone areas near the canal.

(a)

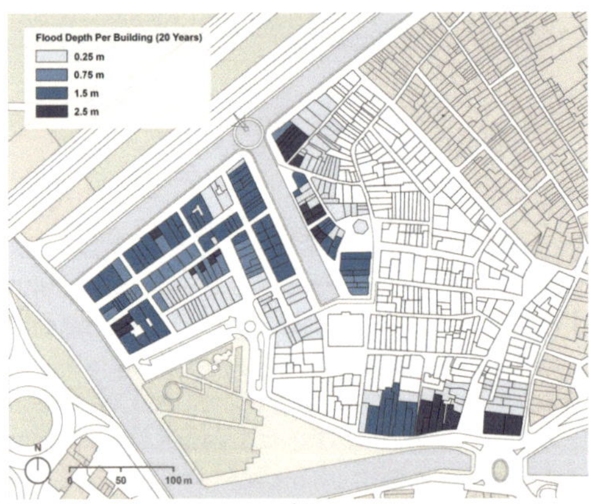

(b)

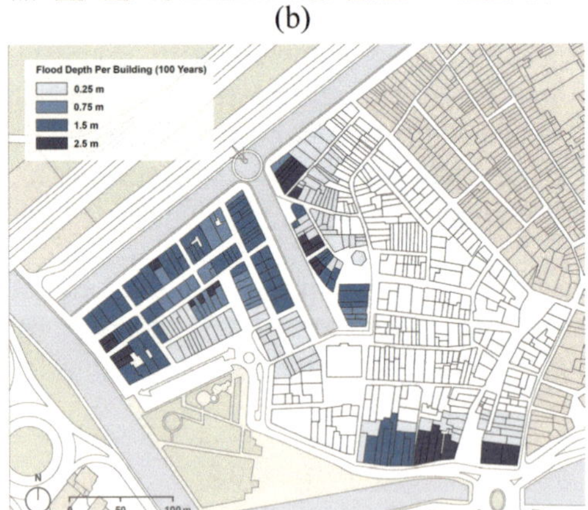

(c)

Figure 10.
Hazard levels: (a) flood depth hazard map for 100 years return period; (b) flood depth per building for 20 years return period; (c) flood depth per building for 100 years return period.

5.3 Risk analysis

By deploying the aforementioned levels of vulnerability and hazard, the flood risk matrix is obtained, as reported in **Table 2**, where the numbers represent the level of risk on a four-point scale.

Hazard level	Hazard indicator range (m/s)	20 years (%)	100 years (%)
Negligible	0.0–0.5	81	80
Low	0.5–1.0	14	15
Moderate	1.0–1.5	5	6
High	1.5–2.0	0	0
Extreme	2.0–2.5	0	0

Table 1.
Distribution of hazard levels for Aveiro.

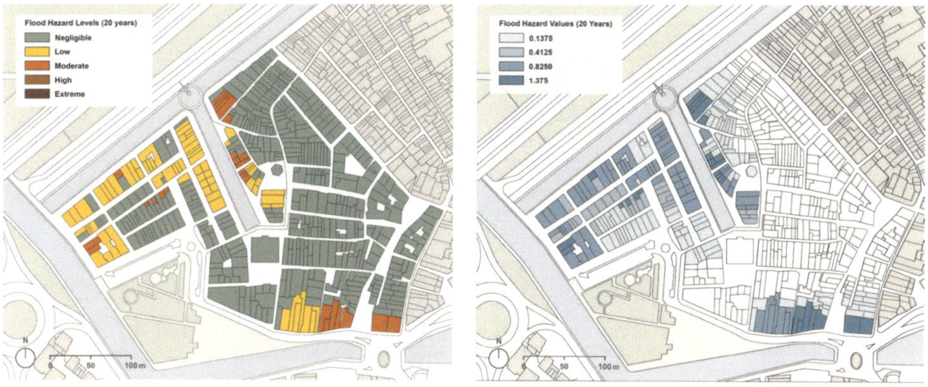

(a) (b)

Figure 11.
Spatial distribution of flood hazard levels (a) and values for the adopted 20-year peak flow scenario (b).

Figure 12 shows the risk maps for the two return periods. The flood risks for the 20-year return period (**Figure 12a**) and 100-year return period (**Figure 12b**) are remarkably similar. In a total of 495 surveyed buildings, 77.37% of them present low flood risk; 12.53 and 12.73% for 100 and 20 years, respectively, moderate; and 10.10 and 9.90% for 100 and 20 years, respectively, high. It is noted that for the 20-year return period, the moderate risk is slightly higher, raising more alarm for a dangerous situation in the near future.

It is observed that the buildings with a significant risk are located closer to the canal; they can be again distinguished into two blocks, and also, a significant number of them are located along one of the main streets, R. João Mendonça. Indeed, the

Flood risk matrix						
Flood risk		**Hazard**				
		Negligible	**Low**	**Moderate**	**High**	**Extreme**
Vulnerability	High	2	3	3	4	4
	Moderate	1	2	3	3	4
	Low	1	1	2	3	3

Table 2.
Flood risk matrix.

(a) (b)

Figure 12.
Spatial distribution of the flood risk results for 20 years (a) and 100 years (b).

Figure 13.
Flood risk per building for a 20-year return period.

buildings likely affected by the flood hazard, in the two return-period scenarios, are also the most vulnerable within the investigated area (**Figure 13**).

Overall, even though a percentage of about 10% of the buildings can be identified as being at high risk in the next decades, it is concluded that the flood risk for the Beira Mar neighborhood is relatively low, and in view of future mitigation strategies, the effort can be concentrated on the buildings located within the identified flood-prone area.

6. Comparison with the Historic City Center of Guimarães

For the purpose of providing a deeper understanding of the integrated approach presented for assessing flood risk in historic city centers, the authors deemed appropriate the development of a comparison between similar case studies. The example selected is the Historic City Center of Guimarães, a UNESCO Heritage Site located in the northern region of Portugal (**Figure 14a**). The case study presented is of particular relevance as it was chosen by Ferreira and Santos in 2020 as a pilot case [21], encompassing 9 blocks with 116 buildings, to implement the same methodology adopted in the present work. The area taken into consideration for the flood risk assessment is located in the "buffer zone" of the declared World Heritage Site area, as shown in **Figure 14b**.

According to Miranda and Ferreira [1], the city of Guimarães has been subjected to strong anthropogenic pressure due to increasing urban and industrial occupation, which originated the current environmental degradation of the Couros river basin as well as the substantial rise in severity of its flooding events. Hereafter, the Historic City Center of Aveiro and Historic City Center of Guimarães case studies will be addressed respectively as "Case A" and "Case B".

For what concerns the exposure and sensitivity components, the comparison between the two case studies is shown in **Figures 15** and **16**. The main difference lies in the exposure component. Case A shows a sharp distinction in the distribution of the classes in comparison to Case B, where the values are more homogeneously arranged. This depends on the wall orientation indicator and reflects the canals' location within the area. The sensitivity, on the other hand, presents a relatively uniform distribution for both cases, never exceeding the range of 40–70, approximately.

Furthermore, the overall flood vulnerability maps are shown for both case studies in **Figure 17**. The range of values for the vulnerability index in Case A never exceeds the 60–70 band, slightly higher than that in Case B, where the maximum registered values range from 50 to 60. However, the main difference lies in the percentage of buildings with recorded vulnerability higher than 30. The quantity is considerably

Figure 14.
The Historic City Centre of Guimarães (a) and identification of the study area (b) [7].

(a) (b)

Figure 15.
Comparison between exposure components of Case A (a) and Case B (b) [21].

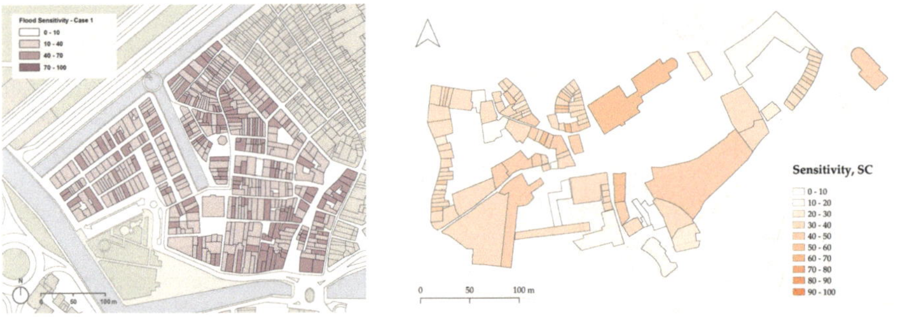

(a) (b)

Figure 16.
Comparison between sensitivity components of Case A (a) and Case B (b) [21].

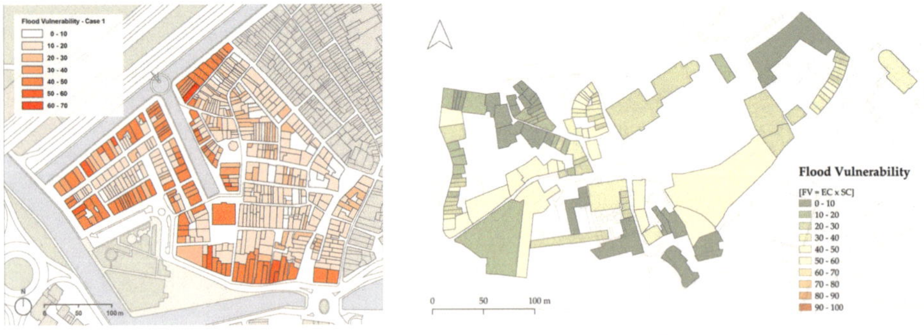

(a) (b)

Figure 17.
Comparison between flood vulnerability of Case A (a) and Case B (b) [21].

greater in Case B, reaching 40% of the 116 evaluated buildings, compared to Case A where the most vulnerable ones only represent 22% of the study area.

As seen, this discrepancy depends on the exposure component: in both cases, the most exposed buildings are also the most vulnerable ones. Even though in Aveiro there appears to be a great number of buildings presenting a higher exposure to the water flow, also considering that the canals almost completely surround the analyzed building stock, in Guimarães the Couros river basin passes through the investigated area. As a consequence, because several of the most sensitive buildings are also located in the central part of the study area, which has a great proximity to the water flow, the final values of vulnerability greatly increase. On the contrary, in Aveiro, the most sensitive buildings are not directly facing the canals.

Hence, it is possible to conclude that when applying the flood vulnerability index to a desired case study, the exposure is indeed a fundamental factor in determining the final flood vulnerability results, but so is the matching cross analysis with the sensitivity component.

7. Conclusions

The present report introduces a thorough study on the flood risk assessment performed in the Historic City Center of Aveiro (Portugal). The study was carried out applying a methodology elaborated by Miranda and Ferreira in 2019 [1], which has been proved to be a useful tool to evaluate the flood vulnerability of built aggregates, with a specific focus on areas distinguished by their heritage value.

A total of about 500 buildings were assessed, including 8 listed heritage assets of local interest. The data collection consisted of an extensive field survey and photographic documentation and was complemented by a remote survey. The spreadsheets of the raw data were post-processed to obtain the vulnerability results, which were defined by combining the exposure component (based on wall orientation) and the sensitivity component (based on heritage status, age, number of stories, condition, and material of buildings). During this process, it was observed that the most sensitive buildings were the ones located along the length of the canals. It is interesting to note that the most sensitive buildings were also revealed to be the most exposed ones.

Following this, the hazard scenario defined for the 20- and 100-year flood was deployed in order to obtain the outputs for flood extent, depth, and velocity. The results that were obtained from the hazard-level classification were deemed as satisfactory, portraying a low level of danger. In the end, the above outputs allowed to perform the final flood risk analysis by merging the hazard scenario with the previously assessed flood vulnerability and thus to acquire an insight on the impact of a future flooding event on the city center. It was concluded that the flood risk for the inspected area remains relatively low, with only 10% of the buildings presenting a high risk in the next decades. The high-risk buildings are located closer to the canal, thus within the identified flood-prone area, and also along one of the main streets of the city center.

Moreover, a comparison was conducted with a portion of the Historic City Center of Guimarães, where a similar approach for the assessment of flood risk was performed. Irrespective of the differences between the two case studies, the influence of the exposure indicator on the vulnerability index of the assets was confirmed.

Finally, it is stressed that the methodology allowed a prompt assessment of the flood risk. Nonetheless, uncertainties arose in the estimation of some parameters, calling for further calibration of the levels and investigations in the study area. Upon this more detailed analysis, the results of the flood risk assessment will be integrated into a decision-making framework for risk mitigation and improvement of the city's resilience.

Acknowledgements

This work was partly financed by FCT/MCTES through national funds (PIDDAC) under the R&D Unit Institute for Sustainability and Innovation in Structural Engineering (ISISE), under reference UIDB/04029/2020. Authors acknowledge all the technical and financial support provided within the framework of the International SAHC Masters Course (www.msc-sahc.org).

Author details

Mayra Alejandra Estrella Núñez[1], Alkmini Firtinidou-Stergiou[1], Margherita Rago[1]*, Chenxin Jonathan Yee[1], Alberto Barontini[1], Tiago Miguel Ferreira[2] and Daniel V. Oliveira[1]

1 Department of Civil Engineering, University of Minho, ISISE, Guimarães, Portugal

2 Department of Geography and Environmental Management, University of the West of England, Bristol, United Kingdom

*Address all correspondence to: arch.margherita.rago@gmail.com

References

[1] Miranda FN, Ferreira TM. A simplified approach for flood vulnerability assessment of historic sites. Natural Hazards. 2019;**96**:713-730. DOI: 10.1007/s11069-018-03565-1

[2] Stone H, D'Ayala D, Wilkinson S. The Use of Emerging Technology in Post-Disaster Reconnaissance Missions. London: The Institution of Structural Engineers; 2017

[3] Stone H, Putrino V, D'Ayala D. Earthquake damage data collection using omnidirectional imagery. Frontiers in Built Environment. 2018;**4**:51. DOI: 10.3389/fbuil.2018.00051

[4] Pittore M, Haas M, Megalooikonomou KG. Riskoriented, bottom-up modeling of building portfolios with faceted taxonomies. Frontiers in Built Environment. 2018;**4**: 41. DOI: 10.3389/fbuil.2018.00041

[5] D'Ayala D et al. Flood vulnerability and risk assessment of urban traditional buildings in a heritage district of Kuala Lumpur, Malaysia. Natural Hazards and Earth System Sciences. 2020;**20**(8): 2221-2241. DOI: 10.5194/nhess-20-2221-2020

[6] Stephenson V, D'Ayala D. A new approach to flood vulnerability assessment for historic buildings in England. Natural Hazards and Earth System Sciences. 2014;**14**(5):1035-1048. DOI: 10.5194/nhess-14-1035-2014

[7] Trizio F, Torrijo FJ, Mileto C, Vegas F. Flood risk in a heritage city: Alzira as a case study. Watermark. 2021;**13**(9):9. DOI: 10.3390/w13091138

[8] Balica SF. Development and Application of Flood Vulnerability Indices for Various Spatial Scales. Delft: UNESCO-IHE Institute for Water Education; 2007

[9] D'Ayala D, et al. ICONHIC 2016, 1st International Conference on Natural Hazards & Infrastructure. Chania, Greece. Assessment of the multi-hazard vulnerability of priority cultural heritage structures in the Philippines. 2016

[10] Drdácký M, Slížková Z. Structural Strategies and Measures Reducing Flood Action on Architectural Heritage. Vol. 2012. Island of Brac. Croatia: WIT Press; 2012. pp. 249-260. DOI: 10.2495/RISK120221

[11] Pickles D. Flooding and Historic Buildings. Historic England: Liverpool University Press; 2020

[12] Snow J. Flood Damage to Traditional Buildings, Historic Scotland. Salisbury Place, Edinburgh: Longmore House; 2014

[13] Vojinovic Z et al. Holistic approach to flood risk assessment in areas with cultural heritage: A practical application in Ayutthaya, Thailand. Natural Hazards. Springer. 2016;**81**(1):589-616. DOI: 10.1007/s11069-015-2098-7

[14] Bigio G, Ochoa MC, Amirtahmasebi R. Climate-resilient, Climate-friendly World Heritage Cities. Washington, DC: World Bank Group; 2014. Available from: https://openknowledge.worldbank.org/handle/10986/19288

[15] Holicky M, Sykora M. Risk Assessment of Heritage Structures Endangered by Fluvial Floods. Milan, Italy: WIT Press; 2010. pp. 205-213. DOI: 10.2495/FRIAR100181

[16] de Pinhal FGR. Aveiro, cidade de água e terra: Um olhar sobre o vernacular no centro histórico. Porto: University of Porto; 2018

[17] Rodrigues MF. A construção civil em Aveiro, 1860–1930: notas para a sua compreensão histórica. In: A indústria Portuense em perspectiva histórica. Actas do colóquio. Aveiro: CLC-FLUP; 1998

[18] Ferreira CF. Vulnerabilidade sísmica do parque edificado na cidade de Aveiro. Aveiro: University of Aveiro; 2008

[19] Fernandes MA. The Adobe Building Culture in Portugal. Icomos; 2013

[20] Lima PMT. Avaliação da Vulnerabilidade e do Risco de Cheia Urbana na Cidade de Aveiro. Aveiro: University of Aveiro; 2018

[21] Ferreira TM, Santos PP. An integrated approach for assessing flood risk in historic city centres. Watermark. 2020;**12**(6):1648. DOI: 10.3390/w12061648

Chapter 3

Threshold Recognition Based on Non-Stationarity of Extreme Rainfall in the Middle and Lower Reaches of the Yangtze River Basin

Yao Wang, Suning Liu, Zhaoqiang Zhou and Haiyun Shi

Abstract

Analyzing the hydrological sequence from the non-stationary characteristics can better understand the responses of changes in extreme rainfall to climate change. Taking the plain area in the middle and lower reaches of the Yangtze River basin (MLRYRB) as the study area, this study adopted a set of extreme rainfall indices and used the Bernaola-Galvan Segmentation Algorithm (BGSA) method to test the non-stationarity of extreme rainfall events. The General Pareto Distribution (GPD) was used to fit extreme rainfall and was calculated to select the optimal threshold of extreme rainfall. In addition, the cross-wavelet technique was used to explore the correlations of extreme rainfall with El Niño-Southern Oscillation (ENSO) and Western Pacific Subtropical High (WPSH) events. The results showed that: (1) extreme rainfall under different thresholds had different non-stationary characteristics; (2) the GPD distribution could well fit the extreme rainfall in the MLRYRB, and 40–60 mm was considered as the suitable optimal threshold by comparing the uncertainty of the return period; and (3) ENSO and WPSH had significant periodic effects on extreme rainfall in the MLRYRB. These findings highlighted the significance of non-stationary assumptions in hydrological frequency analysis, which were of great importance for hydrological forecasting and water conservancy project management.

Keywords: yangtze river, non-stationarity, extreme rainfall, optimal threshold, ENSO, WPSH

1. Introduction

Extreme weather events can change local climate characteristics such as the mean values of temperature and precipitation, as well as their variabilities. As one of the most important climate variables, precipitation, has significant impacts on the hydrological processes and water resources management of a river basin [1]. In the context of climate change, the frequency of extreme weather events has increased, causing more serious natural water-related disasters such as floods and droughts [2–5]. The increase in extreme rainfall always leads to floods, which have remarkable impacts on the local ecology, industry, and social economy. According to the statistics, more than 60% of floods are caused by extreme rainfall since the twentieth century [6, 7]. Therefore, more attention has been paid to studying the importance of extreme rainfall. However, extreme rainfall in different regions can vary greatly in spatial distribution, scope, frequency, duration, and severity [8, 9]. Moreover, the responses to extreme rainfall can vary greatly in different regions. Thus, it is of great importance to investigate the characteristics of extreme rainfall in a designated region.

The traditional hydrological frequency analysis of extreme rainfall and flooding is mainly based on stationary assumptions. However, several studies have shown that extreme rainfall is increasing in many parts of the world due to varying degrees of instability. Li et al. believed that the non-stationarity of the rainfall series could play an important role in the prediction and risk analysis of extreme rainfall [10]. Liu et al. observed that the non-stationarity of extreme rainfall in the Weihe River basin was sensitive to environmental changes [11]. Chen et al. studied the non-stationarity of the maximum daily rainfall in Taiwan and found that the high uncertainty of the non-stationarity of the maximum daily rainfall would lead to a difference in the predicted return period [12]. Beguería et al. established the General Pareto Distribution (GPD) model and found that extreme rainfall in Northwest Spain decreased significantly in winter but increased significantly in summer [13]. Sugahara et al. believed that, under non-stationary conditions, extreme rainfall in Sao Paulo, Brazil, showed a clear increasing trend [14]. Lee et al. found that the peak over threshold (POT)-GPD combined model would be more suitable for predicting future rainfall under non-stationary conditions [15]. Syafrina et al. proposed that the non-stationary extreme rainfall series is more suitable for Generalized Extreme Value Models (GEV) in Sabah [16]. Therefore, it has important significance to study the non-stationarity of extreme rainfall.

Currently, there is no clear definition of extreme rainfall. According to the World Meteorological Organization (WMO), most studies have used percentile rainfall or fixed rainfall as thresholds, such as 95% or 50 mm rainfall [17–20]. However, in a non-stationary state, the difference in the choice of the threshold will generate a large amount of uncertainty, even based on stationary assumptions. Different threshold selections will also have impacts on modeling [7, 21]. Vu and Mishra demonstrated that the models and parameters selected under different thresholds would affect the non-stationarity of extreme rainfall, and a suitable extreme rainfall sequence needed to be selected [22]. Sugahara et al. studied the distribution of extreme rainfall under different thresholds in Sao Paulo and believed that 98% of the daily rainfall value was the most suitable for the extreme rainfall threshold in that region [14]. Liu et al. studied the extreme rainfall in the Weihe River basin and believed that 95% rainfall was suitable [11]. Therefore, it is necessary to study the non-stationarity of extreme rainfall under different thresholds and detect the specific extreme rainfall threshold in a designated region.

Among all, the climatic and non-climatic factors, large-scale climatic patterns, for example, El Niño-Southern Oscillation (ENSO), Pacific Decadal Oscillation (PDO), Western Pacific Subtropical High (WPSH), and Arctic Oscillation (AO), are regarded as the most important factors affecting rainfall [23–27]. Villarini and Denniston showed that ENSO had a significant control effect on extreme rainfall in Australia [28]. Limsakul and Singhruck claimed that PDO was one of the most important factors affecting extreme rainfall changes in Thailand [29]. Fu et al. found that ENSO could affect China's extreme rainfall trends and changes [30]. Zhang and Liu et al. demonstrated that WPSH was one of the main driving factors of the summer extreme rainfall in China [31, 32]. However, previous studies exploring the teleconnections between extreme rainfall and large-scale climatic patterns were usually based on trend analysis and correlation coefficients [33], which could not fully reveal their correlations.

The main purposes of this study are: (1) to divide the middle and lower reaches of the Yangtze River basin (MLRYRB) into sub-regions to perform the non-stationary detection of extreme rainfall, and to identify the non-stationarity of extreme rainfall with different thresholds; (2) to screen out the most suitable threshold range of extreme rainfall in each sub-region based on distribution fitting of extreme rainfall; and (3) to explore the teleconnections between extreme rainfall and large-scale climatic patterns. The main significance of this study is to explore the non-stationary pattern of extreme rainfall in the MLRYRB when the threshold changes. Combined with the non-stationarity of extreme rainfall in different sub-regions, the extreme rainfall threshold range in the region is accurately screened. In addition, the cross-wavelet analysis method will be used to comprehensively explore the relationships between ENSO, WPSH, and extreme rainfall in a different time and frequency domains.

2. Data and methodology

2.1 Study area

The Yangtze River is the longest and largest river in China, which originates in the Tanggula Mountains, flows through 19 provinces from west to east, and finally flows into the East China Sea. The Yangtze River Basin (YRB; 24–35°N, 90–122°E), with a drainage area of about 1.8 million km^2, is located in the subtropical and temperate climate zones dominated by the southeast monsoon. In this study, the MLRYRB (25–34°N, 108–122°E; **Figure 1**) is selected as the study area, which is one of the important water sources and an important economic center in East China. The average annual temperature is between 14 and 18°C, and the average annual precipitation is between 1000 and 1400 mm. Affected by factors such as subtropical monsoons and typhoons, the MLRYRB is one of China's heavy rain-prone areas and areas with the highest flood intensity.

2.2 Data

The data used in this study are obtained from the China Meteorological Data Service Center (http://data.cma.cn/), and there are 62 meteorological stations with the complete sequence of daily rainfall from 1960 to 2020 in the MLRYRB (**Figure 1**). The meteorological stations selected in this study include all types of topography and climate regions in the MLRYRB (**Figure 1**).

To find out the impacts of climate change, the correlations between the extreme rainfall indices and ENSO/WPSH are studied. Here, the Nino3.4 index, which can be obtained from National Center for Atmospheric Research (NCAR) (https://climatedataguide.ucar.edu/climate-data), is used to express the ENSO phenomenon and its changes. The WPSH index, which can be obtained from the National Climate Center of China Meteorological Administration (https://cmdp.ncc-cma.net/Monitoring/cn_stp_wpshp.php), is used to analyze the strong effects of extreme rainfall in the MLRYRB [31, 32].

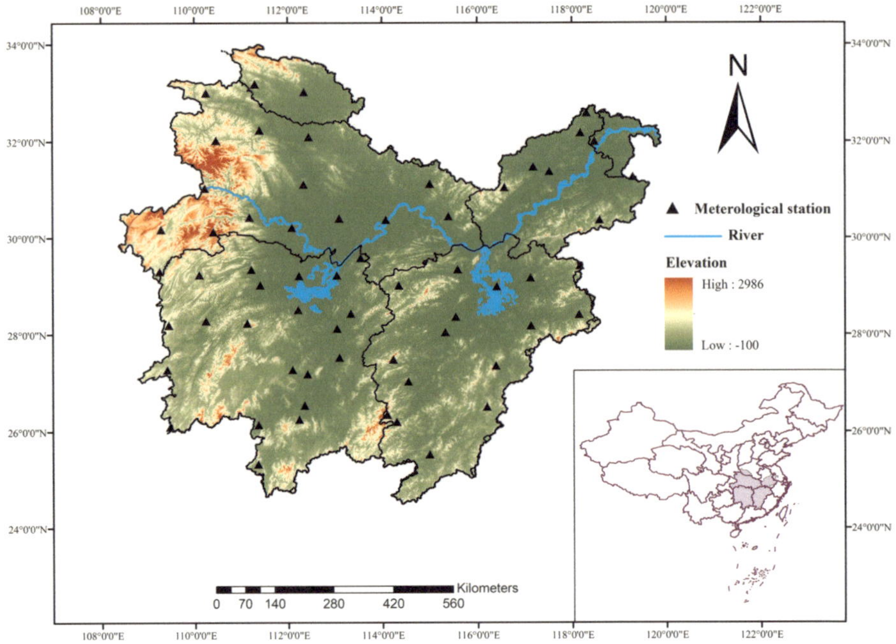

Figure 1.
Location and topography of the MLRYRB in China.

2.3 Methodology

2.3.1 K-mean clustering

The topography of the MLRYRB is complex and the Spatiotemporal distributions of rainfall are uneven due to the influence of the monsoon. Therefore, to reveal the steady changing process and regional differences of extreme rainfall more reasonably, the K-mean clustering method is used to identify similar rainfall characteristics and divide the whole study area into sub-regions. The K-mean clustering analysis is suitable for processing large data sets, which is a commonly used non-hierarchical clustering algorithm with high computational efficiency [11, 34]. This method can reduce the impacts of regional differences on extreme rainfall to a certain extent. For the specific calculation process, please refer to the study of Yang et al. [35].

The results of the K-means clustering analysis vary with the initial cluster centers and the number of selected clusters. Therefore, silhouette coefficient (SC) is usually selected to evaluate whether the number of clusters is reasonable. The value range of SC is from -1 to 1. Generally, the larger the SC value, the better the clustering effect. When the SC value is negative, the clustering result is incorrect. When the average SC value reaches the maximum and the number of negative values is the least, it is the optimal number of clustering. For the specific calculation process, please refer to the study of Liu et al. [19].

2.3.2 Bernaola-Galvan segmentation algorithm

Bernaola-Galvan Segmentation Algorithm (BGSA), a sequence segmentation method proposed by Bernaola-Galvan et al. based on the sliding T-test, is used to detect the mutation points of the extreme rainfall time series. Compared with traditional methods such as the sliding T/F test, the Mann-Kendall (MK) test, and the rank sum test, this method performs better in dealing with highly nonlinear time series [36].

X is a time series with a length of N. One of the segmentation points, i, slides from the left of the sequence to the right in turn. The average values of the left and right parts of the segmentation point are $\mu_1(i)$ and $\mu_2(i)$, and their standard deviations are $s_1(i)$ and $s_2(i)$, respectively. Then, the merged deviation $S_D(i)$ at the split point i can be expressed as Eq. (1).

$$S_D = \left(\frac{(N_1-1)s_1^2+(N_2-1)s_2^2}{N_1+N_2-2}\right)^{1/2} \bullet \left(\frac{1}{N_1}+\frac{1}{N_2}\right)^{\frac{1}{2}} \tag{1}$$

In Eq. (1), N_1 and N_2 represent the sequence length on the left and right sides of the dividing point i, respectively. The difference between the subsequences on both sides of the split point can be represented by the T test statistic $T(i)$ (Eq. (2)).

$$T(i) = \left|\frac{\mu_1(i)-\mu_2(i)}{S_D}\right| \tag{2}$$

In Eq. (2), the T value represents the difference between the subsequences on both sides of the split point. The statistical significance $P(t_{max})$ corresponding to the maximum T is calculated as Eq. (3).

$$P(t_{max}) \approx \left\{1 - I_{\left[\frac{v}{(v+t_{max}^2)}\right]}(\delta v, \delta)\right\}^{\eta} \tag{3}$$

In Eq. (3), $I_x(a,b)$ is an incomplete beta function, where x corresponds to $v/(v+t2\ max)$, a corresponds to δv, and b corresponds to δ. According to the Monte Carlo simulation, η=4.19lnN-11.54, δ=0.40, v=N-2.

If $P(t_{max})$ is greater than or equal to the threshold of $P_0(0.95)$, the difference between the average values is considered to be statistically significant. Then, the time series will be split, and the iteration of the above process will continue until the effective value obtained is less than the threshold or the length of the acquired subsequence is less than the minimum length l_0 ($l_0 \geq 25$). On the contrary, if $P(t_{max})$ is less than $P_0(0.95)$, the sequence will not be split.

2.3.3 Distribution fitting

In the identification of regional extreme rainfall, the POT method is currently used to screen the rainfall sequence, and then the parametric method is used to fit the filtered extreme rainfall sequence. The statistical distribution models used in this study for extreme rainfall research are the GPD and GEV. The distribution functions of GPD and GEV are as follows.

For the GPD in Eq. (4), k is the shape parameter, β is the threshold, and α is the scale parameter.

$$F(x) = \frac{1}{\alpha}\left[1+k\left(\frac{x-\beta}{\alpha}\right)\right]^{-1-\frac{1}{k}} \quad k \neq 0, \beta \leq x \leq \frac{\alpha}{k} \tag{4}$$

For the GEV in Eq. (5), μ is location parameter, α is scale parameter, and k is shape parameter.

$$F(x) = \frac{1}{\alpha}\exp\left[-\left(1+k\left(\frac{x-\mu}{\alpha}\right)\right)^{-\frac{1}{k}}\right]\left(1+k\left(\frac{x-\mu}{\alpha}\right)\right)^{-1-\frac{1}{k}} k \neq 0 \quad (5)$$

Klomogorov-Smirnov test (KS-test) is used to compare the results of different distributions. The larger the p value calculated by the KS-test, the better the fitting result. The standard uses a confidence level of 0.05 (p=0.05). After obtaining the distribution parameters, the corresponding return period in different years is Eq. (6).

$$R = \beta + \frac{\alpha}{k}\left(1-(yT)^{-k}\right) k \neq 0 \quad (6)$$

In Eq. (6), R is the rainfall value corresponding to the return period; y is the number of rainfalls exceeding the threshold in one year, taking the multi-year average value; T is the year of the return period; β is the threshold; α is the scale parameter; and k is the shape parameter.

2.3.4 Cross wavelet analysis

The cross wavelet analysis proposed by Hudgins et al. is an effective tool family for studying the correlation of time series [37]. Cross wavelet transform (CWT) combines wavelet transform and cross spectrum analysis, and can display two sets of time-correlated time-domain sequences. For two time series X and Y, their cross wavelet transform can be defined as Eq. (7).

$$W_n^{xy}(s) = W_n^x(s)W_n^{y*}(s) \quad (7)$$

In Eq. (7), $Wy^*n(s)$ is the complex conjugate of $Wy\,n(s)$, and s is the time lag. The cross wavelet power spectrum is defined as $|Wxy\,n(s)|$, and its value indicates the degree of correlation between the two time series.

For two time series X and Y, the expected spectra are $Px\,k$ and $Py\,k$, then the cross wavelet power spectrum distribution is expressed as Eq. (8).

$$D\left\{\frac{W_n^x(s)W_n^{y*}(s)}{\sigma_x\sigma_y}\right\} = \frac{z_v(p)}{v}\sqrt{P_k^x P_k^y} \quad (8)$$

In Eq. (8), σ_x and σ_y are the standard deviations of the time series X and Y, respectively. z_v (p) is the confidence level related to the probability p, and v is the degree of freedom. The calculation program of cross wavelet analysis can be found in the study of Torrence and Compo [38], and the code can be downloaded from http://noc.ac.uk/using-science/crosswavelet-wavelet-coherence.

3. Results

3.1 Classification of different sub-regions

The MLRYRB has different climatic characteristics, and different sub-regions can receive different influences from monsoon and large-scale climatic patterns. Therefore, it is necessary to determine the areas with the same changing characteristics for division. In this study, it is considered that the variations of extreme rainfall in different regions will be affected by the differences in regional characteristics. Therefore, the K-mean clustering was used to divide sub-regions with the same rainfall characteristics according to rainfall, terrain, longitude, and latitude, and the SC value was used to evaluate the results of clustering. When the SC reaches the maximum, the result is considered to be optimal. The numbers of clusters from 2 to 6 were calculated, and the corresponding SC values are shown in **Table 1**. When the number of clusters was 5, the SC value was the largest (i.e., 0.54). So, the MLRYRB could be divided into five sub-regions (**Figure 2**). Li et al. used Hierarchical Climate Regionalization to study the regionalization of the MLRYRB [39], and their results also showed that the division of five sub-regions was reasonable. After the clustering was completed, the homogeneity test was carried out on the rainfall sequences of the rain gauge stations in each sub-region. Then, the regional attribution of the boundary rain gauge stations was adjusted to ensure that the rainfall sequences of the meteorological stations in each sub-region could pass the homogeneity test.

3.2 Non-stationarity of extreme rainfall with different thresholds

Extreme rainfall events increased with intensified urbanization in the MLRYRB. Previous studies have shown that non-stationary characteristics were found in various

K	2	3	4	5	6
SC	0.268	0.41	0.47	0.54	0.52

Table 1.
Silhouette coefficient (SC) information for the K-means clustering analysis.

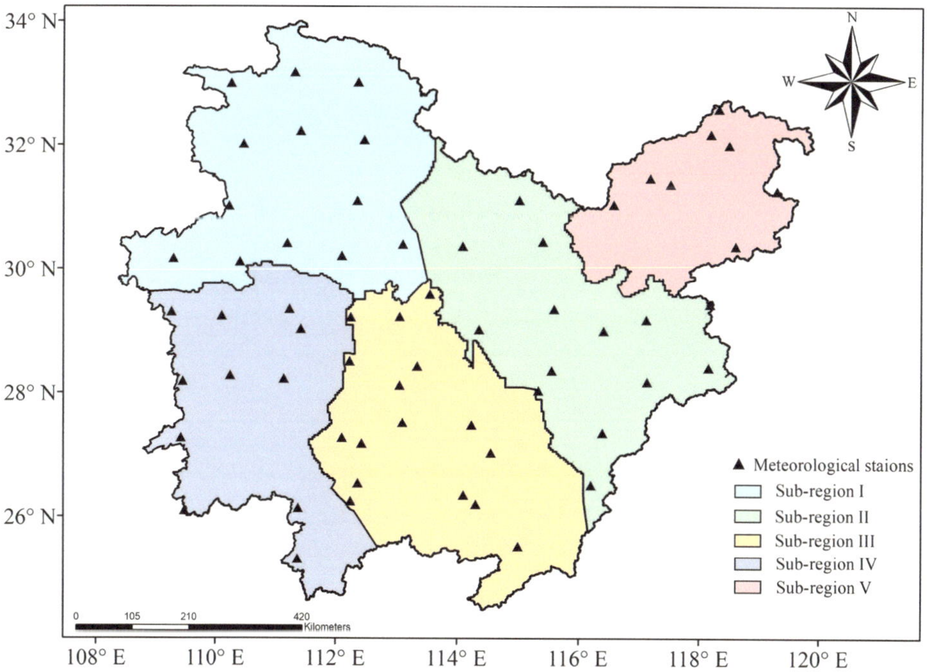

Figure 2.
Five sub-regions with similar dry and wet characteristics in the MLRYRB.

extreme rainfall indices in the MLRYRB, indicating that the assumption of the stability of extreme rainfall would be no longer applicable [10, 40]. Some studies have shown that the extreme rainfall models under different thresholds will generate different non-stationary characteristics [14, 41]. The non-stationarity of rainfall will also affect the analysis of the runoff series [42]. So, the non-stationary detection of extreme rainfall events under different thresholds becomes particularly important.

Figure 3 shows the distributions of the breakpoints of the extreme rainfall events under different thresholds for the five sub-regions, respectively. For Sub-region I, the breakpoints were mainly distributed in 1978, 1980, and 1994. When the threshold changed from 40 to 55 mm, the breakpoints were in 1983, 1986, and 1989. When the threshold was greater than 52 mm, the breakpoint began to appear in 2014, and disappeared when the threshold reached 74 mm. When the threshold was greater than 75 mm, the non-stationarity of extreme rainfall in Sub-region I disappeared. Compared with the whole MLRYRB, changing years in Sub-region I were significantly different. The breakpoints in Sub-region I were sensitive within the threshold of 40–60 mm, while the extreme rainfall events exceeding 75 mm were stable. The situation of Sub-region II was almost consistent with that of the whole MLRYRB, and the difference was that the breakpoints were not continuous with the change of the threshold in 1968. At the same time, when the threshold was less than 50 mm, the breakpoints between 1980 and 2000 were messy. When the threshold was greater than 75 mm, the breakpoints from 1960 to 1980 were still chaotic, which illustrated that the extreme rainfall events fluctuated between 1960 and 1980 in Sub-region II. The breakpoints in Sub-region III changed steadily along with the threshold, and the breakpoints were concentrated around 2000, 1993, and 1970, which was consistent with the overall changing trend. The breakpoints in Sub-region IV varied chaotically along with the threshold, and the breakpoints were relatively stable around 1968, 1970, and 2013; however, they were chaotic between 1970 and 2000, showing that the extreme rainfall events in Sub-region IV fluctuated greatly. The breakpoints of the extreme rainfall sequence in Sub-region V fluctuated relatively steadily along with the threshold, which were concentrated around 2012, 1992, and 1968.

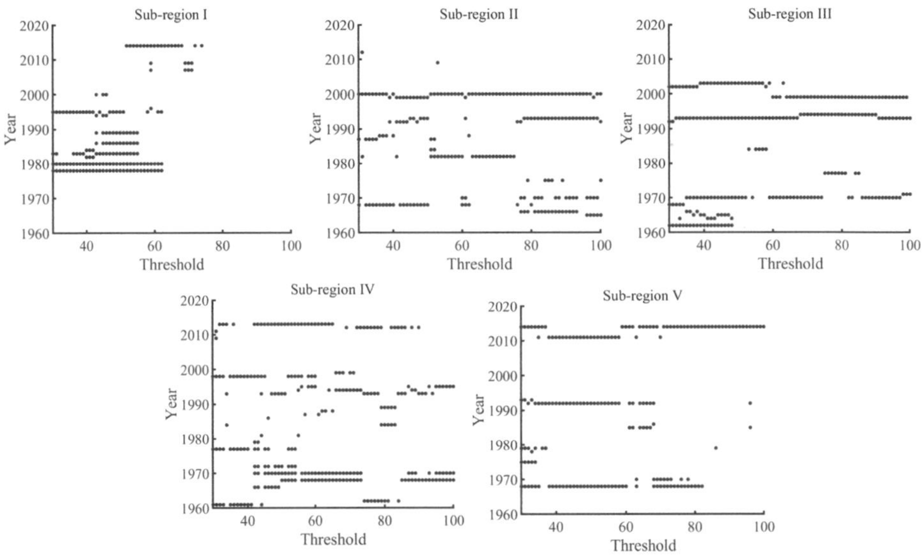

Figure 3.
Year distribution of breakpoints under different thresholds in the five sub-regions of the MLRYRB.

When the threshold was greater than 70 mm, the breakpoint in 1992 disappeared, and when the threshold was greater than 85 mm, the breakpoint in 1968 disappeared. These results showed that the extreme rainfall events with the rainfall over 85 mm were stable. Overall, breakpoints varied widely between subregions. The selection of extreme rainfall thresholds in the five sub-regions of the MLRYRB would have certain impacts on the non-stationarity of extreme rainfall series, and the non-stationarity of extreme rainfall events corresponding to different thresholds would be quite different. This will affect the judgment of environmental factors of hydrological sequence and extreme rainfall.

3.3 Fitting results with different thresholds

In **Figure 4**, it reveals the fitted GPD of the extreme rainfall sequences selected by different thresholds in the five sub-regions of the MLRYRB. Since the p-values of the KS-test fitted by the GEV were all less than 0.05, they are not shown in this figure. For Sub-region I, when the threshold was greater than 40 mm, the p-value increased significantly, and then decreased significantly when the threshold was greater than 50-60 mm. When the threshold was greater than 60 mm, the p-value gradually increased. For Sub-region II, the fitting results were relatively good when the threshold was between 30 and 40 mm, the fitting results were better after the threshold was greater than 60 mm, and the fitting results fluctuated greatly when the threshold was between 40 and 60 mm. For Sub-region III, when the threshold was within 40–50 mm and greater than 70 mm, the fitting results were better. However, when the threshold was between 50 and 70 mm, the fitting results had larger deviations. For Sub-region IV, when the threshold was between 40 and 65 mm, the fitting results were good but fluctuated greatly. After the threshold was greater than 60 mm, the fitting results were not satisfactory. For Sub-region V, when the threshold was within 60–70 mm, the fitting results were the best; however, the fitting results had a large fluctuation in the whole interval of 0–100 mm.

It is shown that variation ranges of different design return periods in the five sub-regions of the MLRYRB in **Figure 5**. For different design return periods, the trends in these five sub-regions were consistent. When the return period increased from 5 to 100 years, the corresponding extreme rainfall values showed increasing trends. When

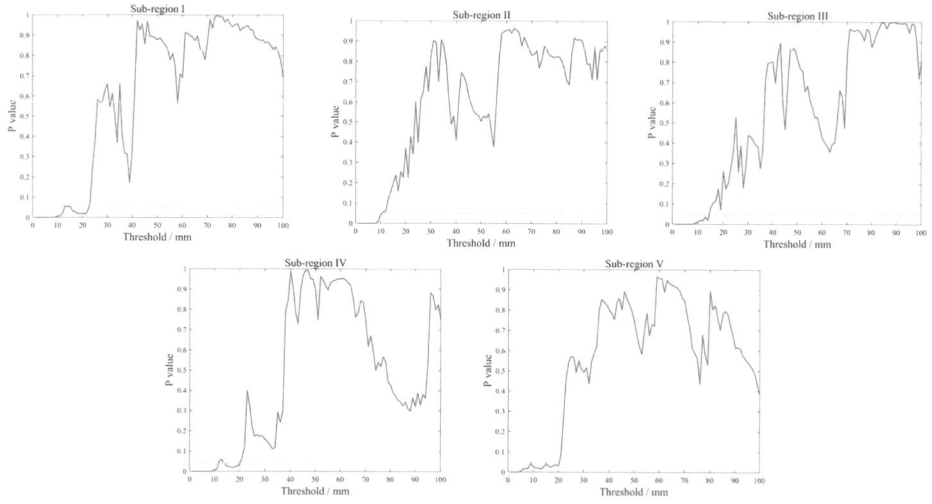

Figure 4.
P-value of KS-test of the GPD fitting with different thresholds in the five sub-regions of the MLRYRB.

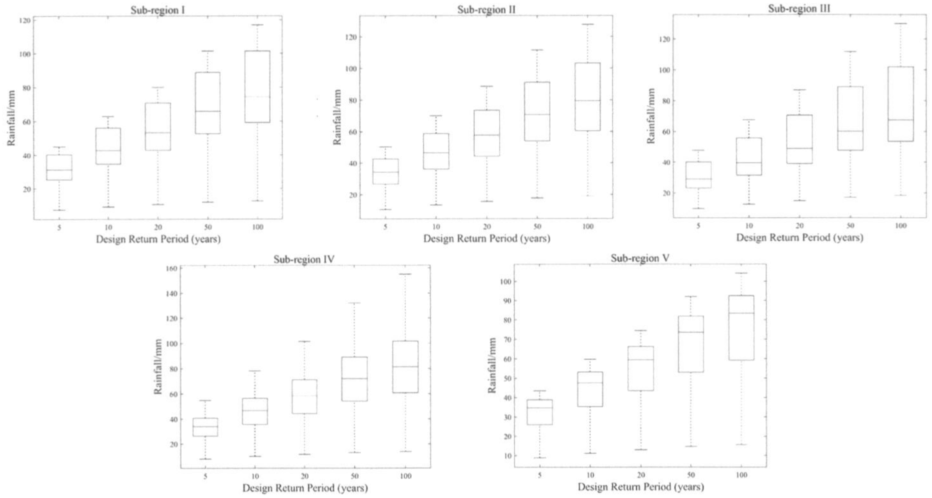

Figure 5.
Variation ranges of different design return periods in the five sub-regions of the MLRYRB.

the threshold was less than 20 mm, different levels of extreme rainfall could not be well differentiated. The rainfall of the 5-year return period in different sub-regions had similar rainfall. The median rainfall of the 10- and 20-year return periods in Sub-regions I, II, III, and IV all had similar performances between 40 and 60 mm, while the threshold in Sub-region V was between 50 and 60 mm. The threshold in Sub-region V was slightly higher than those in the other four sub-regions, which might be related to the geographical location of Sub-region V being closer to the coast. For the median rainfall of the 50-year and 100-year return period, Sub-regions I, II, and IV performed similarly with the threshold between 60 and 80 mm. The threshold was lower than 60 mm in Sub-region III and greater than 80 mm in Sub-region V. In terms of uncertainty for different design return periods, Sub-regions II and III behaved similarly, while Sub-region V had the highest uncertainty and Sub-region IV had the lowest uncertainty.

3.4 Correlations between large-scale climatic patterns and extreme rainfall

Large-scale climatic patterns can potentially affect the non-stationarity of extreme hydrological and meteorological events [43], so correlation analysis of large-scale climatic patterns and extreme rainfall events can help to analyze the causes of extreme rainfall non-stationarity. This study used the CWT to analyze the correlation of the maximum daily rainfall of the year (RX1day) with ENSO and WPSH in the MLRYRB and its five sub-regions (**Figures 6** and 7), as well as their resonance frequency and phase shift in the time-frequency domain.

Figure 6 illustrates the cross wavelet spectrum of RX1day and ENSO for each sub-region. In Sub-region I, there was a periodic signal of 3–5 years during 1968–1976, a periodic signal of 2–4 years during 1994–1996, and a periodic signal of 5–8 years as well as a periodic signal of 9–10 years during 1994–2001. In Sub-region II, RX1day and ENSO showed a significant negative correlation. There was a periodical signal of 2–3 years during 1964—1971, a periodical signal of about 4 years during 1996–2000, and a signal of about 12 years during 2000–2006. In Sub-region III, there was only a

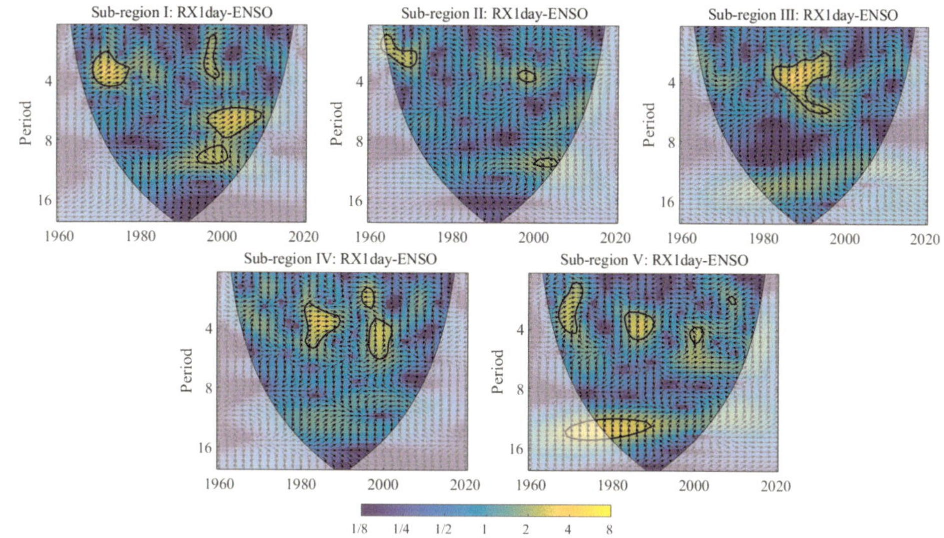

Figure 6.
Cross wavelet power spectra of RX1day and ENSO in the five sub-regions of the MLRYRB. Note: the noise is shown with a thick outline for the 95% significance confidence level, and the relative phase relationship is shown with the arrows. The area enclosed by the black cone outline is considered significant. The phase angle represents the relationship between the two variables. When the phase angle points to right, it indicates a positive correlation between the two variables. When the phase angle points to left, it represents two variables have a negative correlation.

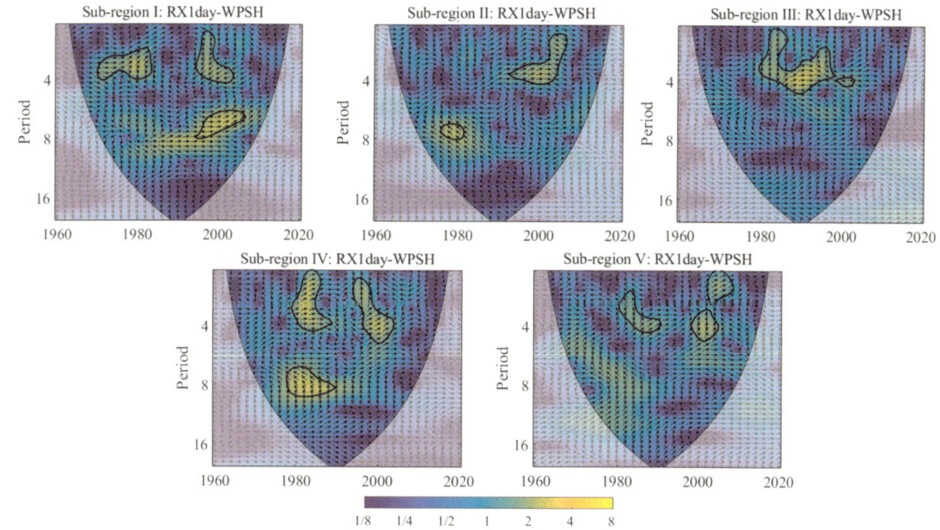

Figure 7.
Cross wavelet power spectra of RX1day and WPSH in the five sub-regions of the MLRYRB.

periodic signal of 2–7 years during 1982–1996. In Sub-region IV, there were periodic signals of 3–5 years, 2–3 years, and 4–6 years distributed during 1980–1990, 1996–1998, and 1997–2000, respectively. Except for some short-period signals of less than 5 years in Sub-region V, the long-period signals of 13–15 years from 1968 to 1990 was noticeable. In general, RX1day in the MLRYRB had a significant positive correlation with ENSO. RX1day in different sub-regions had relatively similar responses to the changes in ENSO. However, in Sub-region II, there was a significant negative correlation between RX1day and ENSO, which was inconsistent with those in other sub-regions. This indicates that the impacts of ENSO on the extreme rainfall of the MLRYRB needs further study.

The cross wavelet spectrums of RX1day and WPSH for each sub-region are displayed in **Figure** 7. In Sub-region I, there were periodic signals of 3–4 years, 2–4 years, and 6–8 years during 1970–1984, 1994–2004, and 1994–2016, respectively. In Sub-region II, there was a negatively correlated periodic signal of 7–8 years from 1976 to 1981 and a positively correlated periodic signal of 2–4 years during 1990–2006. In Sub-region III, there was only one periodic signal during 1978–2003 with 1–5 years. In Sub-region IV, there were two positively correlated periodic signals of 1–5 years during 1978-1988 and 1994-2004, and there was also a negatively correlated periodic signal of 6–10 years during 1976–1990. In Sub-region V, there were periodic signals of 2–4 years, 3–5 years, and 1–2 years during 1979–1989, 1996–2004, and 2002–2008, respectively. Most of the signals showed an obvious positive correlation, but there were also some negative correlation signals. Most of these negative correlation signals appeared before 1990. Moreover, during the same time period, different signal cycles showed the opposite correlations, indicating that WPSH had different effects on RX1day at different times.

4. Discussion

Figure 8 shows the BGSA test results of RX1day in the five sub-regions of the MLRYRB. The thin blue line represents the RX1day sequence, the thick line represents the average RX1day sequence, and the red line represents the split year. Since the BGSA method is based on the non-stationary detection of the mean value, the variation of RX1day will affect the non-stationarity of the extreme rainfall series to a certain extent. It can be considered that the non-stationarity for extreme rainfall series at a given threshold should be similar to the non-stationarity for the maximum rainfall series. After comparing the non-stationarity of the extreme rainfall series screened by different thresholds and RX1day series, the extreme rainfall threshold obtained would have a certain reference. Comparing the non-stationary split point years of extreme rainfall series under different thresholds, the extreme rainfall thresholds in the MLRYRB should be between 40 and 60 mm.

After selecting a more reasonable threshold range, this study compared the non-stationarity and the p-value of the extreme rainfall series under different thresholds. When the extreme rainfall series tends to be relatively stable, the corresponding p-value will also respond higher. Combined with the threshold uncertainty under different return periods, when the threshold was selected below 40 mm, the extreme rainfall series were in a strong non-stationary state, the fitting results were relatively poor, and the calculated design return years could not be distinguished very well. Therefore, below 40 mm was not a suitable threshold of extreme rainfall to screen the rainfall sequence, which is also consistent with the conclusion drawn by the BGSA method. When the rainfall threshold was greater than 60 mm, the p value in Sub-regions II, IV, and V begins to decline. For Sub-regions I and III, although the fitting effect was the best when the threshold was greater than 60 mm, the corresponding return period was not within a reasonable range. Therefore, based on the above findings, the reasonable range of the extreme rainfall threshold of each sub-region in the MLRYRB should be between 40 and 60 mm with the largest p value. Therefore, the extreme rainfall thresholds of the five sub-basins were 45, 60, 50, 45, and 60 mm, respectively. It is found that the

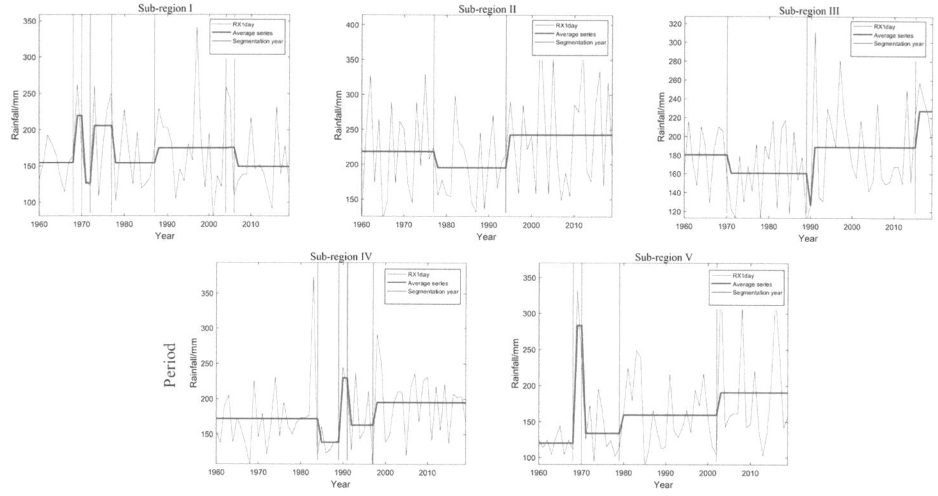

Figure 8.
BGSA test results of RX1day in the five sub-regions of the MLRYRB.

extreme rainfall thresholds of the sub-regions close to the coast were higher than those far from the coast, which might be due to that the sub-regions close to the coast are more susceptible to typhoons. Regarding the whole MLRYRB, the thresholds of 40 and 60 mm corresponds to the 95% quantile and 99% quantile of the daily rainfall. So, the range of 40–60 mm is reasonable when screening and calculating extreme rainfall and different thresholds can be selected for the screening of extreme rainfall sequences for different computing requirements. Due to climate change, the definition of regional extreme rainfall is of great significance for precipitation forecasting and flood protection. The method presented in this study can judge the extreme rainfall thresholds of regions under changing environments and obtain the most suitable thresholds. When the threshold of extreme rainfall is low, it often does not work; when the threshold of extreme rainfall is high, rainfall forecasting tends to increase the risk of local flood disasters. Therefore, the extreme rainfall threshold with regional characteristics is also more reasonable. Moreover, this study adopts a generalized method and has no regional limitations, which means that it will be applicable to other regions and future climate model simulations. For different regions and different environmental backgrounds, the choice of extreme rainfall needs to consider the local specific conditions.

5. Conclusions

This study conducted the threshold recognition based on non-stationarity of extreme rainfall in the MLRYRB. The main conclusions can be summarized as follows:

First, five sub-regions with similar rainfall characteristics were identified. Using the BGSA method to detect the non-stationarity of extreme rainfall events under different thresholds in each region, it is found that the non-stationarity of extreme rainfall events in the MLRYRB would change with the selection of threshold. When the threshold was selected as 40–60 mm, the non-stationarity of extreme rainfall was similar to the non-stationarity of RX1day series, indicating that 40–60 mm should be a reasonable threshold interval of extreme rainfall.

Second, this study performed distribution fitting on the extreme rainfall screened by different thresholds. The fitting of the General Pareto Distribution (GPD) was much better than that of the Generalized Extreme Value Models (GEV) regardless of the threshold selection. Then, the GPD with different design return periods was calculated, and the uncertainty of the threshold in Sub-region V was slightly larger than those in the other four sub-regions. Combined with the uncertainty of the threshold in each sub-region, the extreme rainfall thresholds in the five sub-regions were 45, 60, 50, 45, and 60 mm, respectively. Moreover, the extreme rainfall thresholds of the sub-regions close to the coast were higher than those far away from the coast.

Third, this study investigated the correlations of extreme rainfall with large-scale climatic patterns and found significant correlations between extreme rainfall and ENSO/WPSH. It is worth noting that WPSH is a large-scale climatic pattern that affects the rainfall in the entire YRB, and the impact of WPSH on the heavy rainfall is significantly stronger than that of ENSO. This suggests that accurate outputs from large-scale climate models can help to improve the extreme rainfall predictions in the MLRYRB.

Acknowledgements

This study was supported by National Natural Science Foundation of China (51909117) and Natural Science Foundation of Shenzhen (JCYJ20210324105014039).

Conflict of interest

The authors declare no conflict of interest.

Author details

Yao Wang[1], Suning Liu[2,3], Zhaoqiang Zhou[1] and Haiyun Shi[1]*

1 Shenzhen Key Laboratory of Precision Measurement and Early Warning Technology for Urban Environmental Health Risks, School of Environmental Science and Engineering, Southern University of Science and Technology, Shenzhen, China

2 Center for Climate Physics, Institute for Basic Science, Busan, Republic of Korea

3 Pusan National University, Busan, Republic of Korea

*Address all correspondence to: shihy@sustech.edu.cn

References

[1] Rao GV, Reddy KV, Srinivasan R, Sridhar V, Umamahesh NV, Pratap D. Spatio-temporal analysis of rainfall extremes in the flood-prone Nagavali and Vamsadhara Basins in eastern India. Weather and Climate Extremes. 2020; **29**:100265

[2] Liu SN, Shi HY, Sivakumar B. Socioeconomic drought under growing population and changing climate: A new index considering the resilience of a regional water resources system. Journal of Geophysical Research: Atmospheres. 2020;**125**(15):e2020JD033005

[3] Shi HY, Chen J, Li TJ, Wang GQ. A new method for estimation of spatially distributed rainfall through merging satellite observations, raingauge records, and terrain digital elevation model data. Journal of Hydro-environment Research. 2020;**28**:1-14

[4] Xu B, Zhong PA, Lu QW, Zhou FL, Huang X, Ma YF, et al. Multiobjective stochastic programming with recourses for real-time flood water conservation of a multireservoir system under uncertain forecasts. Journal of Hydrology. 2020; **590**:125513

[5] Zhou ZQ, Shi HY, Fu Q, Li TX, Gan TY, Liu SN. Assessing spatiotemporal characteristics of drought and its effects on climate-induced yield of maize in Northeast China. Journal of Hydrology. 2020;**588**: 125097

[6] Teegavarapu RS. Floods in a Changing Climate: Extreme Precipitation. Cambridge, UK: Cambridge University Press; 2012

[7] Agilan V, Umamahesh NV, Mujumdar PP. Influence of threshold selection in modeling peaks over threshold based nonstationary extreme rainfall series. Journal of Hydrology. 2021;**593**:125625

[8] Sridhar V, Jin X, Jaksa WTA. Explaining the hydroclimatic variability and change in the Salmon River basin. Climate Dynamics. 2013;**40**(7-8): 1921-1937

[9] Weldegerima TM, Zeleke TT, Birhanu BS, Zaitchik BF, Fetene ZA. Analysis of rainfall trends and its relationship with SST signals in the Lake Tana Basin, Ethiopia. Advances in Meteorology. 2018:5869010

[10] Li HS, Wang D, Singh VP, Wang YK, Wu JF, Wu JC, et al. Non-stationary frequency analysis of annual extreme rainfall volume and intensity using Archimedean copulas: A case study in eastern China. Journal of Hydrology. 2019;**571**:114-131

[11] Liu SY, Huang SZ, Huang Q, Xie YY, Leng GY, Luan JK, et al. Identification of the non-stationarity of extreme precipitation events and correlations with large-scale ocean-atmospheric circulation patterns: A case study in the Wei River Basin, China. Journal of Hydrology. 2017;**548**:184-195

[12] Chen PC, Wang YH, You GJY, Wei CC. Comparison of methods for non-stationary hydrologic frequency analysis: Case study using annual maximum daily precipitation in Taiwan. Journal of Hydrology. 2017;**545**:197-211

[13] Beguería S, Angulo-Martínez M, Vicente-Serrano SM, López-Moreno JI, El-Kenawy A. Assessing trends in extreme precipitation events intensity and magnitude using non-stationary peaks-over-threshold analysis: A case study in northeast Spain from 1930 to

2006. International Journal of Climatology. 2011;**31**(14):2102-2114

[14] Sugahara S, da Rocha RP, Silveira R. Non-stationary frequency analysis of extreme daily rainfall in Sao Paulo, Brazil. International Journal of Climatology: A Journal of the Royal Meteorological Society. 2009;**29**(9): 1339-1349

[15] Lee O, Sim I, Kim S. Application of the non-stationary peak-over-threshold methods for deriving rainfall extremes from temperature projections. Journal of Hydrology. 2020;**585**:124318

[16] Syafrina AH, Norzaida A, Ain JJ. Stationary and nonstationary generalized extreme value models for monthly maximum rainfall in Sabah. Journal of Physics: Conference Series. 2019; **1366**(1):012106

[17] Ashcroft L, Karoly DJ, Dowdy AJ. Historical extreme rainfall events in southeastern Australia. Weather and Climate Extremes. 2019;**25**: 100210-100210

[18] Saha U, Singh T, Sharma P, Das Gupta M, Prasad VS. Deciphering the extreme rainfall scenario over Indian landmass using satellite observations, reanalysis and model forecast: Case studies. Atmospheric Research. 2020; **240**:104943

[19] Liu BJ, Chen XH, Chen JF, Chen XH. Impacts of different threshold definition methods on analyzing temporal-spatial features of extreme precipitation in the Pearl River Basin. Stochastic Environmental Research and Risk Assessment. 2017;**31**(5):1241-1252

[20] Grimm AM, Tedeschi RG. ENSO and extreme rainfall events in South America. Journal of Climate. 2008;**22**(7): 1589-1609

[21] Beguería S. Uncertainties in partial duration series modelling of extremes related to the choice of the threshold value. Journal of Hydrology. 2005;**303** (1-4):215-230

[22] Vu TM, Mishra AK. Non-stationary frequency analysis of the recent extreme precipitation events in the United States. Journal of Hydrology. 2019;**575**:999-1010

[23] Ward PJ, Kummu M, Lall U. Flood frequencies and durations and their response to El Niño Southern Oscillation: Global analysis. Journal of Hydrology. 2005;**539**:358-378

[24] Wi S, Valdés JB, Steinschneider S, Kim TW. Non-stationary frequency analysis of extreme precipitation in South Korea using peaks-over-threshold and annual maxima. Stochastic Environmental Research and Risk Assessment. 2016;**30**(2):583-606

[25] Villafuerte MQ, Matsumoto J, Kubota H. Changes in extreme rainfall in the Philippines (1911–2010) linked to global mean temperature and ENSO. International Journal of Climatology. 2015;**35**(8):2033-2044

[26] Lestari S, Hamada JI, Syamsudin F, Sunaryo MJ, Yamanaka MD. ENSO influences on rainfall extremes around Sulawesi and Maluku Islands in the Eastern Indonesian maritime continent. Sola. 2016;**12**:37-41

[27] Sukhatme J, Venugopal V. Waxing and waning of observed extreme annual tropical rainfall. Quarterly Journal of the Royal Meteorological Society. 2016; **142**(694):102-107

[28] Villarini G, Denniston RF. Contribution of tropical cyclones to extreme rainfall in Australia. International Journal of Climatology. 2016;**36**(2):1019-1025

[29] Limsakul A, Singhruck P. Long-term trends and variability of total and extreme precipitation in Thailand. Atmospheric Research. 2016;**169**:301-317

[30] Fu GB, Yu JJ, Yu XB, Ouyang RL, Zhang YC, Wang P, et al. Temporal variation of extreme rainfall events in China, 1961–2009. Journal of Hydrology. 2013;**487**:48-59

[31] Zhang C. Moisture sources for precipitation in Southwest China in summer and the changes during the extreme droughts of 2006 and 2011. Journal of Hydrology. 2020;**591**: 125333

[32] Liu YY, Liang P, Sun Y. The Asian Summer Monsoon: Characteristics, Variability, Teleconnections and Projection. Cambridge, MA: Elsevier; 2019. p. 237

[33] Huang SZ, Huang Q, Chang JX, Zhu YL, Leng GY, Xing L. Drought structure based on a nonparametric multivariate standardized drought index across the Yellow River basin. China. Journal of Hydrology. 2015;**530**:127-136

[34] Zhou ZQ, Shi HY, Fu Q, Ding YB, Li TX, Wang Y, et al. Characteristics of propagation from meteorological drought to hydrological drought in the Pearl River Basin. Journal of Geophysical Research: Atmospheres. 2021;**126**(4): e2020JD033959

[35] Yang Y, Gan TY, Tan XZ. Spatiotemporal changes of drought characteristics and their dynamic drivers in Canada. Atmospheric Research. 2020; **232**:104695

[36] Bernaola-Galván P, Ivanov PC, Amaral LAN, Stanley HE. Scale invariance in the nonstationary of human heart rate. Physical Review Letters. 2001;**87**(16):168105

[37] Hudgins L, Friehe CA, Mayer ME. Wavelet transforms and atmospheric turbulence. Physical Review Letters. 1993;**71**(20):3279

[38] Torrence C, Compo GP. A practical guide to wavelet analysis. Bulletin of the American Meteorological Society. 1998; **79**(1):61-78

[39] Li M, Zhao RX, Wang GW, Chai XR. Precipitation divisions in the middle and lower reaches of the Yangtze River and their meteorological drought time changes. Resources and Environment in the Yangtze Basin. 2020;**12**:2719-2726 (In Chinese)

[40] Song ZH, Xia J, She DX, Zhang LP, Hu C, Zhao L. The development of a nonstationary standardized precipitation index using climate covariates: A case study in the middle and lower reaches of Yangtze River Basin. China. Journal of Hydrology. 2020;**588**:125115

[41] Strupczewski WG, Singh VP, Feluch W. Non-stationary approach to at-site flood frequency modelling I. Maximum likelihood estimation. Journal of Hydrology. 2001;**248**(1-4):123-142

[42] Chang CM, Yeh HD, Chuang MH. Spectral analysis of temporal variability of nonlinear and nonstationary rainfall-runoff processes. Journal of Hydrology. 2019;**575**:1301-1307

[43] Jain S, Lall U. Floods in a changing climate: Does the past represent the future? Water Resources Research. 2001; **37**(12):3193-3205

Chapter 4

On the Use of the Ensemble Kalman Filter for Torrential Rainfall Forecasts

Yasumitsu Maejima

Abstract

Torrential rainfall is a threat to modern human society. To prevent severe disasters by the torrential rains, it is an essential to accurate the numerical weather prediction. This article reports an effort to improve torrential rainfall forecasts by the Ensemble Kalman Filter based on the recent studies. Two series of numerical experiments are reported in this chapter. One is a dense surface observation data assimilation for a disastrous rainfall event caused by active rainband maintained for a long time. Although an experiment with a conventional observation data set represents the rainband, the significant dislocation and the underestimated precipitation amount are found. By contrast, dense surface data assimilation contributes to improve both the location and surface precipitation amount of the rainband. The other is the rapid-update high-resolution experiment with every 30-second Phased Array Weather Radar (PAWR) data for an isolated convective system associated with a local torrential rain. The representation of this event is completely missed without the PAWR data, whereas the active convection is well represented including fine three-dimensional structure by PAWR data assimilation. Throughout these studies, the data assimilation by Ensemble Kalman Filter has a large positive impact on the forecasts for torrential rainfall events.

Keywords: ensemble Kalman filter, numerical weather prediction, convective scale meteorology, torrential rainfall, rainfall forecasts

1. Introduction

The torrential rainfall is becoming one of the most significant treats in worldwide human society year by year. It often brings severe disaster in social infrastructure or human lives. In Eastern Asia, the disasters caused by heavy rainfalls are repeated as every year. In Japan, many rainfall events have brought severe disasters especially in summer season. Among them, in September 2015, active rainband maintained in eastern part of Japan for several days, and severe flood happened. Eventually, it caused 20 fatalities and more than 7000 house damages [1]. In July 2018, a record-breaking and catastrophic torrential rainfall occurred over a wide area of western Japan. According to a report by Fire and Disaster Management Agency, Japan, it caused 263 fatalities, 484 injuries, and more than 30,000 home damages [2]. In another case, the monsoon front so-called Baiu front (in Japanese) or Meiyu front (in Chinese) remained stagnant for long time in July 2020, large amounts of water vapor tended to flow into the Japan islands. This caused record-breaking rainfall over a wide area of Japan. Especially in Kumamoto prefecture located southwest part of Japan, floods of unprecedented magnitude occurred [3–6]. Besides these events, sudden local torrential rainfalls also often happened in summer season. On September 11, 2014, an isolated cumulonimbus suddenly developed within 10 minutes, and over 50 mm h^{-1} intense rainfall was observed, even though it was sunny until then. Recently, even in Europe, where climatological annual rainfall amount is relatively low, less than 1000 mm per year, the occurrence of torrential rains is not unusual. For example, the ECMWF reported extreme rain in Germany and Belgium in July 2021 [7].

To reduce disasters caused by severe rainfall events, a development of an accurate numerical weather prediction (NWP) system is essential. To improve the forecast accuracy, more accurate initial conditions are needed. Accurate initial values can be obtained through advanced initial value creation methods that maximize the use of information from observations, and data assimilation plays a central part in it. In modern NWP system, variational method and Kalman Filter, especially Ensemble Kalman filter [8], are widely applied. Although both schemes have their advantages and disadvantages, Ensemble Kalman Filter is generally superior to implementation and maintenance [9]. The reason why Ensemble Kalman filter rather than the traditional Kalman filter is applied in the NWP is the computational cost. It is quite difficult to apply the traditional Kalman filter technique due to the huge computational cost mainly in the estimation of error covariance matrix. The NWP models has very large degree of freedom, and it usually consists over 10^7 grid points and over 10 variables. On the other hand, in Ensemble Kalman Filter, since error covariance matrix is estimated from the ensembles of the order of 10–100, it is expected to significantly reduce computational costs. In this chapter, the author reports recent studies of improving the torrential rainfall forecasts. In Section 2, an impact of dense surface data assimilation on the forecast of band-shaped rainfall zone is presented. In Sections 3 and 4, the importance of rapid-update data assimilation for a local torrential rainfall having the order of 10 km or less and a practical application for the forecast are presented. In Section 5, the conclusion is described.

2. The dense surface observation data assimilation for a forecast of a torrential rainband

2.1 Introduction

In the torrential rainfall event in Japan, band-shaped rainfall areas are often observed. In Japan, it is widely called "Senjo-Kousui-Tai," an example of translation is "linear rainband." According to an explanation by JMA, this type of rainband is explained as follows: a rainfall area with intense precipitation extending in 50–300 km length and 20–50 km width, created by organized cumulonimbus clouds, which are formed by a series of developing rain clouds passing or lingering in the same location for several hours [10]. As the development of observational networks, torrential rainbands are frequently captured, and they are one of the most essential factors in severe meteorological disasters. In this section, a record-breaking rainfall event in Kanto, Tohoku region, which is located in eastern Japan, in September 2015, is focused. In this event, the southerly winds from Typhoon No. 18, made landfall in the central region of mainland of Japan and moved into the Sea of Japan. At about the same time, the southeasterly winds from Typhoon No. 17 moved northward in Pacific Ocean over the eastern Honshu region. This moist air mass flowed into the Kanto and southern Tohoku regions. The air mass maintained a pronounced convergence zone for a long time, resulting in the development of a series of torrential rain bands. Eventually, the accumulated precipitation since the beginning of the rainfall reached 647.5 mm, more than twice the monthly precipitation normal for September. In addition, 16 of the JMA's 1300 surface observation stations (Automated Meteorological Data Acquisition System; AMeDAS) recorded the highest maximum 24-hour precipitation in observation history [11, 12].

The active rain band itself had a scale of several hundred kilometers; however, the peak precipitation, which directly related to severe disasters, was concentrated in a narrow region. From the viewpoint of heavy rainfall forecasting and disaster prevention, it is desirable not only to improve the accuracy of simulation of a 100 km-scale phenomenon, but also to be able to simulate particularly intense local precipitation with pinpoint accuracy in terms of both location and amount. In this section, as a modern NWP study, a series of data assimilation experiments with dense surface observation data for the heavy rainfall event in September 2015 are reported. The surface observation data comes from the Environmental Sensor Network provided by NTT DoCoMo, which is a major mobile phone company in Japan, and it has about 4000 stations throughout Japan. This observation network has approximately 5-km special intervals and 1-minute temporal resolution at minimum. This study focuses on the impact of this dense surface observation data on the forecast for the rainfall.

2.2 Experimental design

This series of experiments was performed by an NWP system called SCALE-LETKF [13]. It combines the Local Ensemble Transform Kalman Filter (LETKF) [14] with a regional numerical model Scalable Computing for Advanced Library and Environment (SCALE) [15].

The workflow is visualized in **Figure 1**. Firstly, 18-km-mesh model with 271x243x50 grid points, 6-hour-update, 50-member SCALE-LETKF was performed from 0000 JST on September 7, 2015, to 0000 JST on September 10, 2015. Initial condition of ensemble mean and boundary condition came from National Center for Environmental Prediction, Global Forecast System (NCEP GFS). The initial perturbation for generating initial ensemble state came from the perturbation components of NCEP Global Ensemble Forecast System (GEFS). The assimilated observation data are NCEP PREPBUFR. It consists conventional observation by airplanes, ships, buoys, satellites, surface stations, radiosondes. The locations and elements of the observation are summarized in **Figure 2** and **Table 1**; however, some observation systems such as the orbital satellites, airplanes, and ships are not stationary. Thus, it needs to pay attention that the locations in **Figure 2** are just an example at a single time. Next, 4-km-mesh, hourly update data assimilation experiments were performed. The initial ensemble states were interpolated from 18-km-mesh data at 0000 JST on September 7, 2015. The hourly boundary conditions came from a simple forecast at 18-km resolution initialized at 0000 JST on September 7, 2015. To investigate impacts of rich surface observation data, two experiments were performed. One experiment only assimilates NCEP PREPBUFR (CTRL). Although NCEP PREPBUFR is delivered every 6 hours, it has been divided into hourly segments with reference to time stamps for the hourly update LETKF cycles. The other assimilates rich surface observation

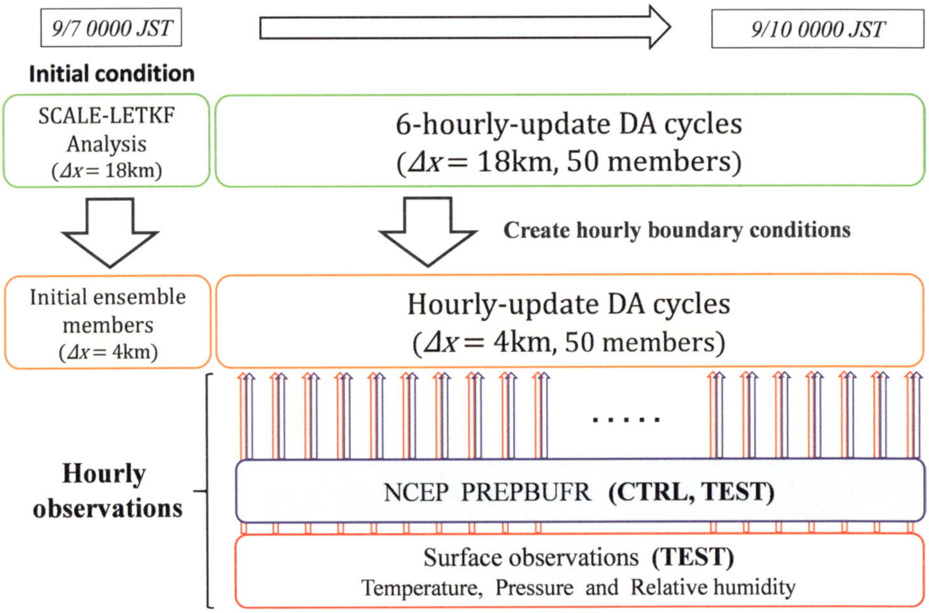

Figure 1.
The workflow of the data assimilation experiments. NCEP PREPBUFR was input both in CTRL and TEST, and surface observation was input in TEST only.

data in addition to NCEP PREPBUFR (TEST). The surface data have been interpolated to the nearest numerical model grid point by bilinear scheme in horizontal and been applied an altitude correction using moist-adiabatic lapse rate in vertical as an observation operator for data assimilation. The surface stations observe wind velocity, pressure, temperature, relative humidity, rainfall amount, solar radiation, and detection of precipitation. In this experiment, wind velocity, pressure, temperature, and relative humidity were assimilated.

2.3 Result and discussion

2.3.1 Result of the analyses

Firstly, the difference of relative humidity between CTRL and TEST after the first data assimilation cycle is shown in **Figure 3**. The impact of surface data is clearly appeared, and positive values are widespread. In this case, numerical model underestimated the surface relative humidity; in other words, it predicted drier condition. However, owing to the dense surface data assimilation, the lower relative humidity was well modified.

Next, the rain mixing ratios [g kg^{-1}] at the lowest model level (approximately 20-m altitude) at 1200JST and 1800 JST on September 9, 2015, are shown in **Figure 4** This period corresponds to observing the intense precipitation by gauge observation around the disaster area. In both cases, peak rainfall mixing ratios reached the order of 1 g kg^{-1}, indicating that intense precipitation is occurring. However, in a quantitative comparison, particularly intense precipitation areas above 2 g kg^{-1} are only found in TEST. Focusing on the distribution of rain mixing ratio, the CTRL shows

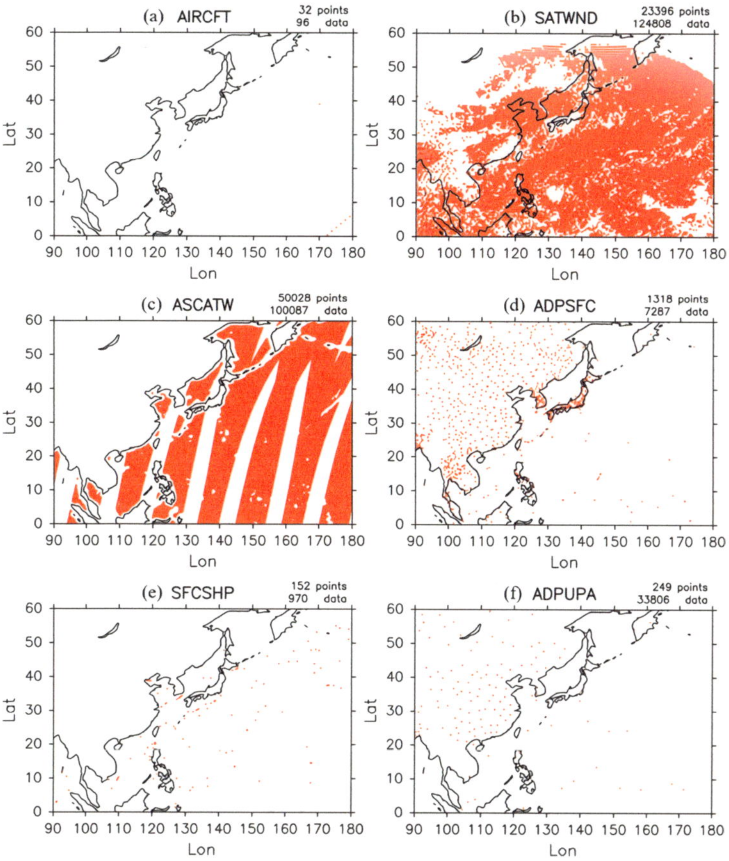

Figure 2.
An example of the observation sites of NCEP PREPBUFR. The valid time is 1200 UTC on December 1, 2019. (a) Aircraft observations, (b) satellite-assigned atmospheric tracking winds, (c) satellite scattered measured winds, (d) surface observations, (e) ship and buoy observations, and (f) radiosonde observations, respectively.

(a) AIRCFT	Aircraft observations	Horizontal wind [m s^{-1}], Temperature [K], Pressure [hPa] Specific humidity [kg kg^{-1}], Geopotential height [m]
(b) SATWIND	Satellite-assigned atmospheric tracking winds	Horizontal wind [m s^{-1}], Pressure [hPa]
(c) ASCATW	Satellite scattered measured winds	Horizontal wind [m s^{-1}], Geopotential height [m]
(d) ADPSFC	Surface observations	Horizontal wind [m s^{-1}], Temperature [K], Pressure [hPa] Specific humidity [kg kg^{-1}]
(e) SFCSHP	Ship and buoy observations	Horizontal wind [m s^{-1}], Temperature [K], Pressure [hPa] Geopotential height [m]
(f) ADPUPA	Radiosonde observations	Horizontal wind [m s^{-1}], Temperature [K], Pressure [hPa] Specific humidity [kg kg^{-1}], Geopotential height [m]

Table 1.
Observation elements in the NCEP PREPBUFR. The elements enumerate all the data stored in the files, including those that are diagnostically calculated.

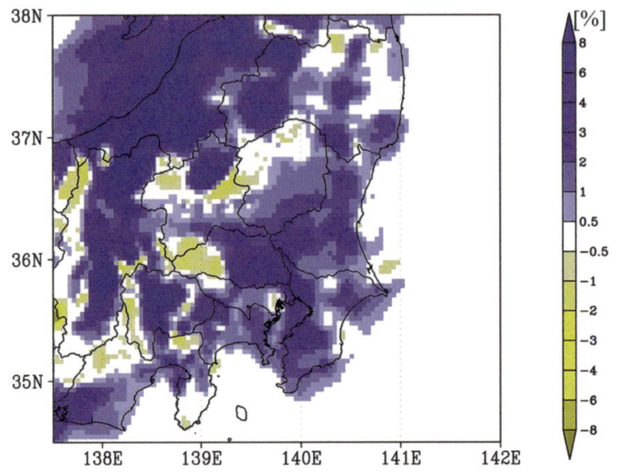

Figure 3.
The difference of surface relative humidity between TEST and CTRL after the first data assimilation cycle at 0100 JST, September 7.

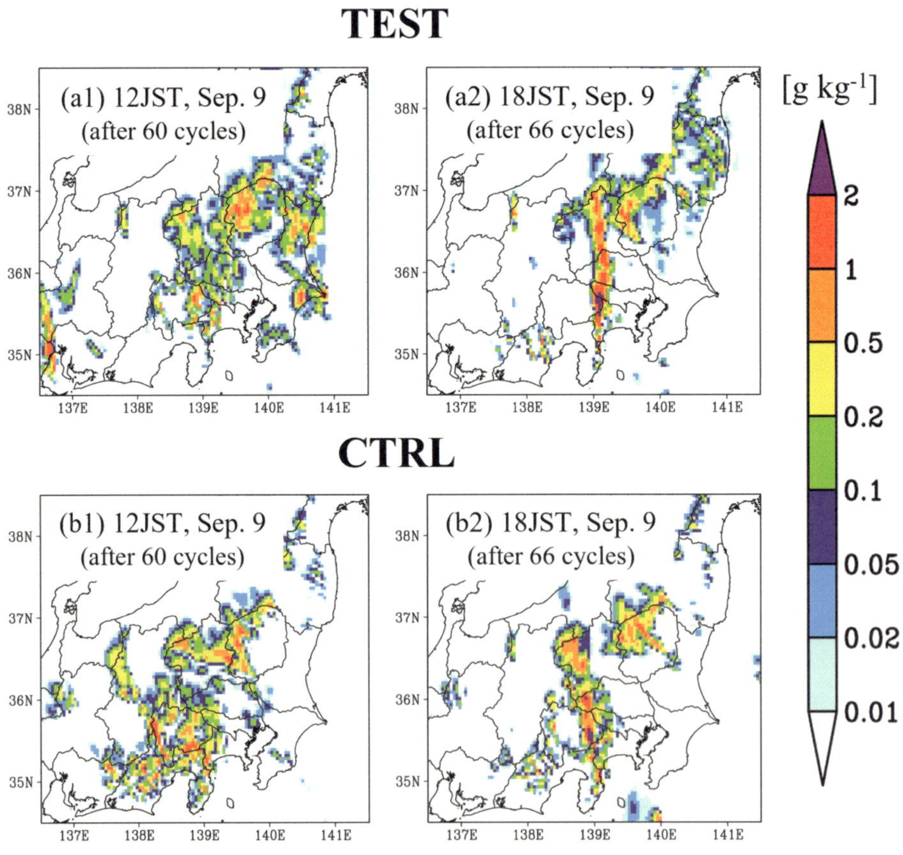

Figure 4.
Rain mixing ratio of the ensemble mean analyses at (a1, b1) 1200 JST, September 9, and (a2, b2) 1800 JST, September 9. Top and bottom rows correspond to TEST and CTRL, respectively.

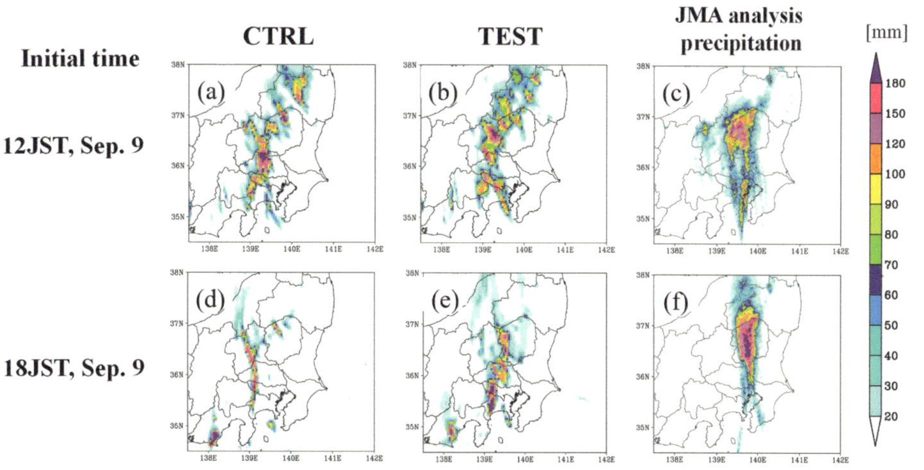

Figure 5.
Six-hour accumulated precipitation amount [mm] in the forecasts initialized at (a) 1200 JST, September 9 and (b) 1800 JST, September 9.

a northwesterly tilt of the precipitation zone. In CTRL, the path of typhoon No. 18, which brought rich moisture into the disaster area, had westerly bias, so the moist air mass from the typhoon No. 18 flowed into the different area from the actual observation. In TEST, the issue of shifting to the west remains; however, precipitation area extended in meridional direction, and it has consistency with the actual phenomenon. The dense surface data assimilation contributed to reproduce the distribution and amount of rain mixing ratio closer to reality than CTRL.

2.3.2 Forecast experiment

Here, 6-hour forecasts initialized at 1200 JST and 1800 JST were performed. The initial conditions come from the analyses of the ensemble mean of CTRL and TEST at each initial time. **Figure 5** shows 6-hour accumulated precipitation amount [mm] of the forecasts. For reference, analysis precipitation, which is the best estimation of surface precipitation intensity at 1-km resolution by JMA, is also shown. In CTRL, regardless of the initial time, the precipitation area is shifted about 100 km to the west compared with the JMA analyses. Moreover, accumulated precipitation amount in CTRL was underestimated. At 1800 JST, even though the peak precipitation amount was 80% level of the JMA analysis, a quite narrow rainband was represented. In TEST, dislocation of the precipitation areas was improved and quantitatively consistent with JMA analyses.

3. The rapid-update Kalman filtering for a torrential rainfall event

3.1 Introduction

This section reports a study of rapid-update ensemble Kalman filtering for a torrential rainfall event on September 11, 2014, in Kobe city, which locates in the western part of Japan and provides a discussion of an impact on a torrential rainfall forecast.

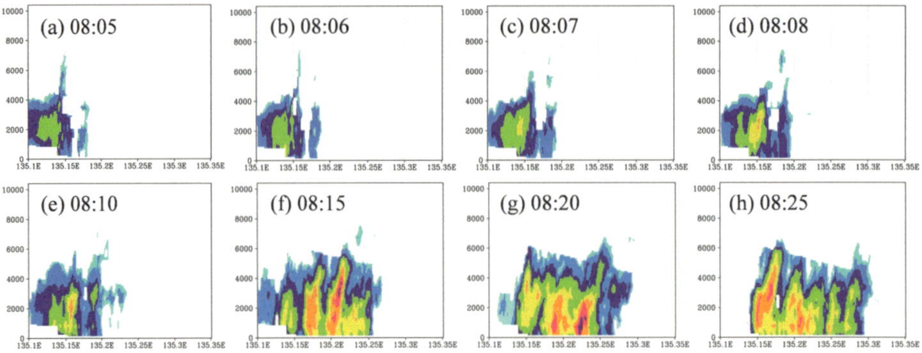

Figure 6.
Zonal-vertical cross section of the radar reflectivity [dBZ] of an isolated convective system generated around Kobe city on September 11, 2014. (a) 0805 JST, (b) 0806 JST, (c) 0807 JST, (d) 0808 JST, (e) 0810 JST, (f) 0815 JST, (g) 0820 JST, (h) 0825 JST.

In this case, an isolated convection system suddenly occurred near Kobe City. **Figure 6** shows a developing process of the cumulonimbus, which was captured by an innovative meteorological observation instrument called phased array weather radar (PAWR) located at Osaka University, Suita city, Osaka, Japan [16]. The PAWR enables to observe a range of 60-km radius with approximately 100-m special resolution and 30-second time resolution. Although it has not been used for the operational weather forecast yet, it has been utilized for a lot of research of the NWP [5, 17–19]. Also in this event, the PAWR at Osaka University [16] well captured the convective initiation and developing process with very high special and temporal resolution. The PAWR observed the first echo at 0758 JST, and its intensity was at most 1 mm h^{-1} in terms of surface precipitation. However, at 0808 JST only 10 minutes later, it was up to 50 mm h^{-1}. Despite the occurrence of torrential rain, the operational forecast completely missed this event. It is because the operational NWP system has not been designed for such an isolated convective system so far. Although this event did not bring the large-scale disaster such as a flood or a landslide, it was a case of social impact. Many people mourned on SNS that they were hit by a sudden heavy rain during the commuting time because the weather forecast on the morning of the day did not broadcast the rainfall forecast. The failure prediction could be attributed that both horizontal resolution and data assimilation window of the operational NWP systems did not match for the local-scale phenomena. To fix this issue, fine resolution and rapid data assimilation cycles are required in the NWP system. Toward forecasting such a torrential rainfall, this study tried to perform 30-second-update, 100-m-mesh data assimilation experiment. This study aims to investigate the impact of such a high-resolution and rapid-update NWP system on the forecast of an isolated convective system. The descriptions in this section are based on a part of a published journal article [17].

3.2 Experimental settings

Here, general settings of a series of data assimilation experiment are described. For details, see the original paper [17].

This series of experiments used the Japan Meteorological Agency nonhydrostatic model (JMA-NHM) [20–22] implementing LETKF [14]. This NWP system is called NHM-LETKF [23, 24]. JMA-NHM was used as the operational mesoscale NWP model in JMA until February 2017. The numerical experiment took quintuple downscaling strategy. First, 15-km-mesh, 6-hour-update, 100-member NHM-LETKF was performed for 10 days from 0900 JST September 1, 2014. The initial and boundary conditions were JMA global spectrum model (GSM) initialized at 0900 JST September 1, 2014, and the perturbations among the ensemble members came from forecast data by JMA global ensemble prediction system (GEPS) initialized at the same time.

Next, 5-km-mesh and 1-km mesh ensemble simulations initialized by the result of the15-km-mesh run were performed. The result of 1-km-mesh ensemble run provided the initial and boundary conditions for the following data assimilation experiments. In the main experiments, 100-m-mesh, 30-second-update LETKF with every 30-second PAWR observation data was cycled (100 M). To discuss the dependence of horizontal resolution, 1-km-mesh, 30-second-update LETKF was also cycled (1 K). These main experiments were performed from 08:00 JST to 08:30 JST on September 11, 2014, and they provided every 30-second, 100-m-mesh, 100-member analysis data. For reference, a similar experiment but without inputting any observation was performed (NO-DA).

Following the series of data assimilation experiments, 30-mitute forecasts initialized by the ensemble-mean analyses at 0830 JST were performed to evaluate the impacts on the torrential rainfall forecasts.

3.3 Result and discussion

To evaluate the impact of every 30-second PAWR data assimilation on the precipitation, the ensemble-mean analyses of the radar reflectivity (dBZ) at 0830 JST corresponding to the last data assimilation cycle are shown in top panels of **Figure** 7(a0-d0). Although the numerical model does not predict the radar reflectivity directory, it can be calculated by the observation operator applied in the PAWR data assimilation experiment as follows:

$$\mathrm{dBZ} = 10 \times \log_{10}\left\{\left(2.53\times 10^{4}\right)(\rho QR)^{1.84} + \left(3.48\times 10^{4}\right)(\rho QS)^{1.66} + \left(8.18\times 10^{4}\right)(\rho QG)^{1.50}\right\} \quad (1)$$

where ρ, QR, QS, and QG are air density [kg m^{-3}], mixing ratios of rain, snow, and graupel [g kg^{-1}], respectively [25]. The mixing ratio means mass of particles in 1-kg of dry air.

In NO-DA, the entire area is less than 15 dBZ, which is visualized by white shade. The white-colored area can be assumed no precipitation; namely, we hardly find any convective initiations. Both in 1 K and 100 M, intense radar echo corresponding to the precipitation is created, and their locations are good agreement with the PAWR observation. However, the peak intensity of radar echo is significantly different between 1 K and 100 M. In 100 M, the center of the convections shown by over 45 dBZ is clearly found. By contrast, 1 K only shows 36 dBZ at the peak. This underestimation is critical to the reproducibility of the precipitation, because 45 dBZ in 100 M corresponds with over 23 mm h^{-1} of precipitation, whereas the precipitation intensity in 1 K only reaches about quarter level of that in 100 M. The time series of every 30-second analysis in 100 M was visualized in a three-dimensional movie, and it is available on YouTube: https://www.youtube.com/watch?v=s2PgH0mZ7G0 [26].

Below the second panels of **Figure** 7 (a1-a3, b1-b3, c1-c3) show surface precipitation intensity at 0840, 0850, 0900 JST (10-, 20-, 30-minute forecasts) in NO-DA, 1 K, and 100 M, respectively. For verification truth, PAWR observation is shown in parallel (**Figure** 7 d1–d3). While the PAWR observation shows intense echo of over 45 dBZ or more, the 1 K shows less than 35 dBZ. Moreover, in 1 K, the radar reflectivity

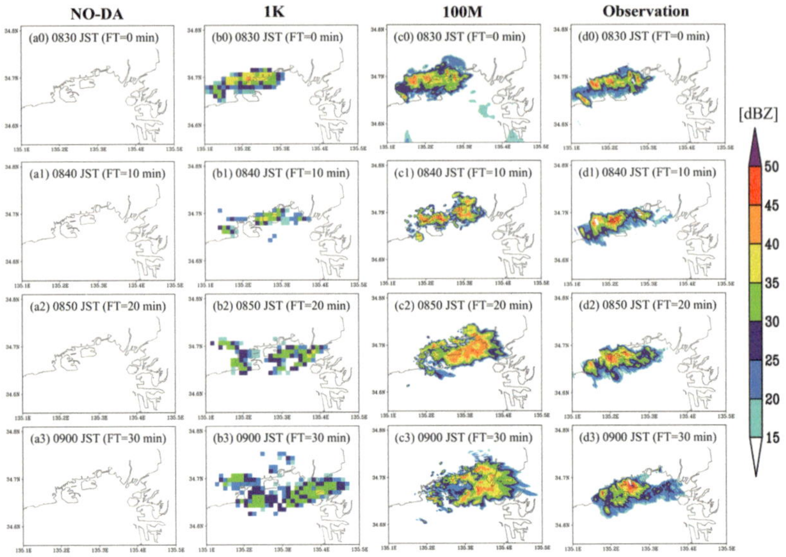

Figure 7.
Radar reflectivity [dBZ] at 2-km elevation for (a0–a3) NO-DA, (b0–b3) 1 K, (c0–c3) 100 M and (d0–d3) PAWR observation. Top, second. Third and bottom rows correspond to 0830 (initial time), 0840 (10-minute forecast), 0850 (20-minute forecast), 0900 JST (30-minute forecast), respectively.

declined as the forecast progressed. In 100 M, both intensity of reflectivity and precipitation area show improvement (**Figure** 7 c1–c3).

Here, one tendency of the forecast should be mentioned. In 100 M, after 20-minute from the start of the forecast (0850 JST), although precipitation intensity is consistent with the PAWR observation, the echo area appears to shift eastward. In 100 M, intense updraft over 20 m s^{-1} was generated and promoted excessive formation of ice crystals. The ice crystals are transported to the east by westerly general winds at about 4 -km level and fall as precipitation particles. That is a causation of the east-shift bias of the rainfall area. Toward performing more accurate forecasts, numerical scheme or model parameters mainly in cloud microphysics should be optimized for simulations with such a high resolution. Also, general simulation designs, such as downscaling strategy, model domains should be reconsidered in near future.

From these comparisons, the high frequency of data assimilation cycles and the very high horizontal resolution contribute to the creation of desirable initial values for forecasting the heavy rainfall. The importance of rapid update data assimilation cycle mentioned in Section 3 has been confirmed by this study. Even though 1-km resolution is fine horizontal grid spacing compared with the operational forecast systems, much higher resolution, which fully enables to resolve the active convections, has large advantage in the analyses and the forecasts.

4. The importance of rapid-update Kalman filtering for the local torrential rainfall

This section provides an explanation of the importance of rapid-update Kalman Filtering for the local torrential rainfall. Here, the "local" refers to a special scale of the order of 10 km or less.

In the first place. The weather system has strong nonlinearity, and the error grows exponentially as the time progresses. Moreover, the weather system has multiscale structure. Large-scale phenomena such as jet stream and have slower error growth, whereas the smaller special scale phenomena have the faster error growth. In torrential rainfalls, cumulus convection plays a main part, and it is the key issue that how to accurately estimate the state of cumulus convection. A relation between special scale and predictive skill is well described in Forecast User Guide by ECMWF [27]. According to **Figure 7(b1)** in the document by ECMWF [27], cumulus convection has 1-km order special scale, and it develops within 10 minutes; thus, fine temporal resolution is required to capture the developing process of the cumulus convection. Recently, sensing technologies have been exploring, and they also have been applied to weather observation instruments, such as radars or satellites. An example of the observation of the cumulus convection has already shown in **Figure 6**. Here, let us see the rapid growth of a convective system in **Figure 6**, carefully. At 0805 JST, only weak radar echo is found. It is a signal of the initiation of rainfall system (**Figure 6a**). At this time, the rainfall intensity is about only 1 mm h^{-1}. The minute-by-minute observation appears to be continuously changing (**Figure 6a–d**). However, this system rapidly evolved within 10 minutes, an active convection having the intense precipitation was formed by 0815 JST (**Figure 6f**).

As mentioned in Section 1, the data assimilation has become an important technique, and Kalman filter is widely applied to convective-scale weather forecasts. The important issue in Ensemble Kalman filter is how to set the length of the data assimilation window. In the first place, Kalman filter assumes linear theory. To apply it to the nonlinear system such as NWP model, we assume that a linear approximation holds within one data assimilation window. So, setting too long assimilation window will destroy the assumption of a linear approximation, resulting in a significant reduction in estimation accuracy. It is repeatedly mentioned that an active convection system, which brings a torrential rainfall, rapidly develops within 10 minutes or so [27] (**Figure 6**). Thus, short-time data assimilation window should be taken in case of applying the ensemble Kalman Filter to torrential rainfall forecast. Here, how long should the data assimilation window be short? Now, let us see **Figure 6** again. Even though the convective cloud is rapid growth system, it can be considered linear growth within a few minutes range. Of course, the shorter the time progress, the more linearity is ensured. From the perspective of Ensemble Kalman Filter, the shorter data assimilation window is taken, the higher accuracy is expected, because the Kalman Filter assumes linear theory. On the other hand, the computational cost will explosively increase by the very short window; however, this issue has been overcome with the innovation of supercomputers. In the previous sections, the author reported a 30-second-update data assimilation experiments for a torrential rainfall event by the previous Japanese flagship supercomputer "K" based on the contents of a published paper. Recently, more powerful supercomputers, such as the current Japanese flagship supercomputer "Fugaku," are operated all over the world. With the development of supercomputers, more accurate NWP system will be performed in operation.

5. Conclusion

This chapter presented the investigations of the forecasts in several torrential rainfall events using Ensemble Kalman Filter. These descriptions are based on my recent studies [17]. In Section 2, I aimed to investigate the impact of dense surface observation data on a hazardous rainfall event in eastern Japan area in September 2015. In this case, a convergence zone extending in meridional direction was maintained by the surrounding wind of two typhoons. It associated to activate the rainband and cause a record-breaking precipitation. So, an impact on the rainband was the primary focus of the study. In this study, two data assimilation experiments were performed. Although the conventional data set (NCEP PREPBUFR) has many kinds of observations, it is difficult to perform precise forecast for the torrential rain. In addition to it, dense surface observation data assimilation contributed to improve forecast accuracy in both rainfall area and amount. Surface observation network is a classical system compared with modern remote sensing instruments. Even though total data size from 4000 stations is the order of 100 kB at one observation time, its data assimilation has certain impact on a torrential rainfall forecast having horizontal scale of the order of 100 km. Sections 3 and 4 showed that the rapid-update data assimilation can be a powerful tool in forecasting a local torrential rainfall through a single case study. In Section 3, a rapid-update ensemble Kalman filtering study for a torrential rainfall event on September 11, 2014, in Kobe city, which locates in the western part of Japan and provides a discussion of an impact on a torrential rainfall forecast. Since the local convective system having 10-km or less special scale grows within 10 minute. The conventional forecast system, which has 1-hour or longer data assimilation cycles and 1-km order horizontal resolution, has hardly resolved an isolated convective system both especially and temporary. To address this issue, 30-second-update, 100-m-mesh experiment was performed. Owing to the innovative NWP system, the torrential rain by an isolated convective system was well represented including its three-dimensional structure, and forecast accuracy was clearly improved. In Section 4, in general, these results were investigated by case studies. As mentioned in Section 3, several issues remained especially in the 100-m-mesh forecast. In order to contribute the results of the research to improve forecasts universally, additional experiments should be conducted and profiled the issues. Moreover, to review the entire parameter settings in Ensemble Kalman Filter is also important.

Acknowledgements

The author thanks the research scientists of data assimilation for the weather forecast, especially the authors of cited papers and websites. Also, the author thanks Ms. Silvia Sabo and Ms. Martina Scerbe, who are the author service managers for a great support of the author's writing activity. In the visualization, free software, the Grid Analysis and Display System (GrADS) developed by Center for Ocean-Land-Atmosphere Studies, Institute of Global Environment and Society, George Mason University, was used.

Author details

Yasumitsu Maejima
RIKEN Center for Computational Science, Kobe, Japan

*Address all correspondence to: yasumitsumaejima@riken.jp

References

[1] A report by Japan Meteorological Agency (in Japanese), 2015, Available from: https://www.jma.go.jp/jma/kishou/books/saigaiji/saigaiji_2015/saigaiji_201501.pdf [Accessed: July 28, 2022]

[2] A report by Fire and Disaster Management Agency, Japan (in Japanese), 2018, Available from: https://www.google.com/url?sa=t&rct=j&q=&esrc=s&source=web&cd=&cad=rja&uact=8&ved=2ahUKEwjb1u3djZv5AhVFCYgKHb6yAmwQFnoECAUQAQ&url=https%3A%2F%2Fwww.fdma.go.jp%2Fdisaster%2Finfo%2Fitems%2Fh30-7_59.pdf&usg=AOvVaw3EEzoeEePYinIOMgXkGDP2 [Accessed: July 28, 2022]

[3] Duc L, Kawabata T, Saito K, Oizumi T. Forecasts of the July 2020 Kyushu heavy rain using a 1000-member ensemble Kalman filter. SOLA. 2021;**17**(17):41-47

[4] Hirockawa Y, Kato T, Araki K, Mashiko W. Characteristics of an extreme rainfall event in Kyushu district, southwestern Japan in early July 2020. SOLA. 2020;**16**:265-270

[5] Maejima Y, Kawabata T, Seko H, Miyoshi T. Observing system simulation experiments of a rich phased array weather radar network covering Kyushu for the July 2020 heavy rainfall event. SOLA;**2022**(18):25-32

[6] Taylor J, Amemiya A, Honda T, Maejima Y, Miyoshi T. Predictability of the July 2020 heavy rainfall with the SCALE-LETKF. SOLA. 2021;**17**:48-56

[7] ECMWF Newsletter No. 169, 2021, https://www.google.com/url?sa=t&rct=j&q=&esrc=s&source=web&cd=&ved=2ahUKEwj275mQkJv5AhWNEogKHdAKAEgQFnoECAUQAQ&url=https%3A%2F%2Fwww.ecmwf.int%2Fsites%2Fdefault%2Ffiles%2Felibrary%2F2021%2F20225-newsletter-no-169-autumn-2021.pdf&usg=AOvVaw02IYtZji-npFQV50xxDsmt [Accessed: July 28, 2022]

[8] Evensen G. The ensemble Kalman filter: Theoretical formulation and practical implementation. Ocean Dynamics. 2003;**53**:343-367

[9] Miyoshi T. A Report on the Workshop on 4D-Var and Ensemble Kalman Filter Inter-Comparisons and Intensive Course on Data Assimilation (in Japanese). Tenki; 2009. pp. 17-24

[10] An official webpage of Japan Meteorological Agency 2022: Available from: https://www.jma.go.jp/jma/kishou/know/yougo_hp/kousui.html, [Accessed: July 28, 2022]

[11] Fujita M, Sato T, Yamada TJ, Kawazoe S, Nakano M, Ito K. Analyses of extreme precipitation associated with the Kinugawa River flood in September 2015 using a large ensemble downscaling experiment. Journal of Meteorology Society of Japan. 2019;**97**(2):387-401

[12] Ushiyama T, Sayama T, Iwami Y. Development of a flood forecasting system using regional ensemble prediction – Application to the Kinugawa flood in 2015. Journal of Japan Society of Civil Engineering. 2019;**73**:124-132

[13] Lien G-Y, Miyoshi T, Nishizawa S, Yashiro H, Adachi SA, Yamaura T, et al., editors. The Near-Real-Time SCALE-LETKF System: A Case of the September 2015 Kanto-Tohoku Heavy Rainfall. SOLA. 2017;**17**:1-6

[14] Hunt BR, Kostelich EJ, Szunyogh I. Efficient data assimilation

for spatiotemporal chaos: A local ensemble transform Kalman filter. Physica D. 2007;**230**:112-126

[15] Nishizawa S, Yashiro H, Sato Y, Miyamoto Y, Tomita H. Influence of grid aspect ratio on planetary boundary layer turbulence in large-eddy simulations. Geoscientific Model Development. 2015;**8**:3393-3419

[16] Owens RG, Hewson TD., ECMWF Forecast User Guide. Reading: ECMWF, 2018, Available from: https://confluence.ecmwf.int/display/FUG/Forecast+User+Guide [Accessed: July 28, 2022]

[17] Ushio T, Wu T, Yoshida S. Review of recent progress in lightning and thunderstorm detection techniques in Asia. Atmospheric Research. 2015;**154**:89-102

[18] Maejima Y, Kunii M, Miyoshi T. 30-second-update 100-m-mesh data assimilation experiments: A sudden local rain case in Kobe on September 11, 2014. SOLA. 2017;**13**:174-180

[19] Maejima Y, Miyoshi T, Kunii M, Seko H, Sato K. Impact of dense and frequent surface observations on 1-minute-update severe rainstorm prediction: A simulation study. Journal of Meteorological Society Japan. 2019;**97**:253-273

[20] Maejima Y, Miyoshi T. Impact of the window length of four-dimensional local ensemble transform Kalman Filter: A Case of Convective Rain Event. SOLA. 2020;**16**:37-42

[21] Saito K, Fujita T, Yamada Y, Ishida J, Kumagai Y, Aranami K, et al. The operational JMA nonhydrostatic mesoscale model. Monthly Weather Review. 2006;**134**:1266-1298

[22] Saito K, Ishida J, Aranami K, Hara T, Segawa T, Narita M, et al. Nonhydrostatic atmospheric models and operational development at JMA. Journal of Meteorological Society Japan. 2007;**85B**:271-304

[23] Saito K. The Japan meteorological agency nonhydrostatic model and its application to operation and research. In Tech. Atmospheric Model Applications. 2012;**2012**:85-110

[24] Miyoshi T, Aranami K. Applying a four-dimensional local ensemble transform Kalman filter (4D-LETKF) to the JMA Nonhydrostatic Model (NHM). SOLA. 2006;**2**:128-131

[25] Kunii M. The 1000-member ensemble Kalman filtering with the JMA nonhydrostatic mesoscale model on the K computer. Journal of Meteorological Society Japan. 2014b;**92**:623-633

[26] Xue M, Tong M, Zhang G. Simultaneous state estimation and attenuation correction for thunderstorms with radar data using an ensemble Kalman filter: Tests with simulated data. Q.J.R. Meteorological Society. 2009;**135**:1409-1423

[27] RIKENchannel on Youtube, 2016: https://www.youtube.com/watch?v=s2PgH0mZ7G0 [Accessed: July 28, 2022]

Chapter 5

Flood Damage on Agricultural Land and Methods for Restoration of Agricultural Soils after Catastrophic Floods in Cold Areas

Trond Knapp Haraldsen

Abstract

Catastrophic floods have large effect on agricultural land both in short and long term. In this chapter, examples of impact of floods of different size in cold regions with glaziers have been presented. The largest floods occur as combination of heavy rainfall and melting and snow and ice in the mountainous areas. Periods of waterlogging by cold running water resulted in decreased yields, but N-fertilization after the soil no longer was water saturated could reduce the yield loss considerably. Although the floods cause severe erosion and sedimentation, results show that it is possible to find measures for reconstruction of the soils with the same productivity as undamaged soils, while the average result was about 85% of the original productivity.

Keywords: agriculture, cereals, erosion, fertilization, flooding, potatoes, restoration, sedimentation, yields

1. Introduction

In recent years, large and catastrophic floods have received much attention. It has been claimed that the large floods occur more often than previously, and the large floods have a more catastrophic pattern. A wilder and wetter climate due to climatic change has been predicted, causing more severe floods with large damage on infrastructure, agricultural soils, loss of lives of animals and humans. In Europe, the number of catastrophic floods was reduced in the period 1870–2016, while the number of humans losing their lives during catastrophic floods increased [1]. A timespan of about 150 years of measurements is too little in order to verify the effects of the largest floods, which may occur once in a millennium or less often. The mechanisms behind the largest floods may be of interest to understand in order to make models for future climatic change.

In many countries, fluvial plains are important agricultural areas with high productivity, but these areas are also the areas that are most exposed to floods. The pattern of Fluvisols [2] with sediments that receive fresh material or have received it in the past and still show stratification indicates areas exposed to floods. Large floods make severe damage on agricultural land by directs damage on crops, erosion, and sedimentation, and different methods for evaluation of flood damage to agriculture have been developed [3]. However, measures for land reclamation and restoration of damaged agricultural areas have sparsely been reported in scientific literature and seem to be absent for areas where combination of snowmelt, melting of glaziers, and heavy precipitation occur.

The aim of this chapter is to show the importance of large floods and the impact on agricultural land both in a historical context and more recent examples with a special focus on restoration and reclamation measures after catastrophic floods in cold regions. The study area is the two largest valleys in Norway; Gudbrandsdalen with the river Gudbrandsdalslågen and Østerdalen with the river Glomma in central eastern Norway, which was severely damaged by a flood in June 1995 [4]. The same area was hit by the largest flood in historical time in July 1789 [5], while the geomorphology of the valleys has been formed by large outburst floods from glacial lakes about 10,000 years before present [6].

2. Material and methods

After the 1995 flood in central eastern Norway, research projects were started in order to document the direct damage of the flood by field experiments at flooded areas. After the direct effects were studied, the effects of different restoration measures were investigated. One of the areas with largest damage after the 1995 flood in the river Glomma was Øksna, north of Elverum in Østerdalen (60°58′ 15″ N, 11°29′ 05″ E). The restoration work took place in the period 1996–1998, and an evaluation of the different measures was done based on studies in a 4 year period, 1999–2002. In total, 11 different measures, including undamaged references, were studied at 26 plots (**Table 1**, **Figure 1**).

Type No.	Damage	Restoration	Texture	Plot No.
1	Erosion crater, >2 m deep	Filled with sand	Loamy fine sand	1, 4, 7
2	Eroded topsoil	Leveling of topsoil	Sandy loam	2, 5, 8
3	Slightly eroded topsoil	Normal tillage	Sandy loam	3, 6, 9
4	Erosion crater, > 2 m deep	Filled with sand, limed sewage sludge mixed in topsoil	Loamy fine sand	10, 11, 12
5	Intact topsoil	Normal tillage	Loamy sand	13, 26
6	Small damage	Leveled, topsoil replaced	Sandy loam	14, 25
7	Erosion crater	Filled with till and sand in bottom, sandy loam as topsoil, sewage sludge mixed in topsoil	Sandy loam	15, 24
8	Small damage	Leveled, topsoil replaced	Sandy loam	16, 23
9	Erosion crater	Filled with sand, sewage sludge mixed in topsoil	Loamy fine sand	17, 22
10	Erosion crater	Filled with till and sand in bottom, sandy loam as topsoil, sewage sludge mixed in topsoil	Sandy loam	18, 21
11	Erosion crater	Filled with till and sand in bottom, sandy loam as topsoil, peat mixed in topsoil (upper 10 cm)	Loamy fine sand	19, 20

Table 1.
Flood damage and restoration measures at Øksna, Elverum commune, Norway.

Figure 1.
Plots for soil and yield investigations at Øksna, Elverum commune, Norway. Photo from 1998 [7].

Two different types of sewage sludge were used as soil conditioner, a limed sewage sludge with pH >10 (Plots 10–12), and a sewage sludge where aluminum salts had been used for coagulation for removal of P from wastewater. The amount of sludge used was about 20 Mg dry matter ha^{-1}, which is the permitted amount according to Norwegian regulation on organic fertilizer.

The yield of cereals, spring barley (*Hordeum vulgare*) and oats (*Avena sativa*), was monitored at the plots in the period 1999–2002. Analysis of variance (ANOVA) was performed in order to document if there were significant crop yield differences between the three groups of treatments: eroded topsoil or intact topsoil, leveled areas, and filled erosion craters. As there were unequal sizes of the groups, the Bonferroni method for multiple comparisons was used ($p < 0.05$).

Soil properties were investigated at plot level, including texture, pH (H_2O), total organic C, representing the topsoil (Ap-horizon). Readily available plant nutrients were determined according to the AL-method [8]. Intact soil cores (100 cm^3) were taken for determination of bulk density and total porosity at nine plots with three cores per plot in the topsoil (5–9 cm depth). Total porosity was measured as water content at saturation, while bulk density was determined after drying the soil cores 24 h at 105°C.

3. Effects on flooding on crops and restoration of flood damage

In the period from 22th May to 2nd June 1995, large part of central eastern Norway was hit by a severe flood, in media called a catastrophic flood. In the period from 22th to 25th May, the temperature rose by 5–10°C causing enormous snowmelt in the mountainous region. It was estimated that 4000 million m^3 snow melted during the period from 25th to 2nd June, which is equivalent to 100 mm precipitation distributed over the whole watershed. In addition, 50–70 mm precipitation as rain hit the central part of the watershed during the days from 28th May to 1st June. This caused a large flood in the rivers Glomma in Østerdalen and the river Gudbrandsdalslågen in Gudbrandsdalen. The maximum observed discharge in Glomma was close to 3100 $m^3 s^{-1}$ at Elverum [4].

During the 1995 flood, about 14,000 hectares agricultural land was flooded, of which about 10,000 hectares along Glomma river in Østerdalen. It has been estimated that 2 million m^3 was eroded by the flood, while the amount of sediments overlying soils constituted of at least 1 million m^3. About 1000 hectares was seriously damaged either by erosion or sedimentation and could not be used without restoration measures [4].

3.1 Crop damage of flooding and effects of fertilization

As the flood in central eastern Norway happened after sowing in the start of the growing season in 1995, it was possible to make research on crop damage and measure if fertilization after the flooding could make reasonable crop yields.

It was found that potatoes had almost total crop failure after waterlogging, even if the period of waterlogging lasted only 1–2 days. Areas with grass for silage and hay production had delayed growth, but recovered after waterlogging and gave reasonable yields [9]. Waterlogging of fields with cereals gave significant yield loss, but the duration of the waterlogged period had less impact (**Table 2**).

As the flooding caused leaching of soluble plant nutrients from the soil and fertilizer, which had been applied in the spring, a series of field experiments were placed

Waterlogged	Solør			Selsvollene		
	Yield kg ha^{-1}	Rel. yield	Protein %	Yield kg ha^{-1}	Rel. yield	Protein %
0 days	620	100	14.1	762	100	13.9
1–2 days	398	64	13.9	627	82	13.7
4–5 days	371	60	13.2			

Table 2.
Influence on waterlogging on yield (kg ha^{-1}, 15% water content) and protein (% of DM) of spring wheat (Triticum aestivum) on fields in Østerdalen (Solør) and Gudbrandsdalen (Selsvollene) in 1995.

at waterlogged locations in Gudbrandsdalen and Østerdalen in order to measure the effect of fertilization after the soil was no longer water saturated [9].

Fertilization with N fertilizer after the fields had recovered sufficient bearing capacity after waterlogging had very good effect on crop yield (**Tables 3** and **4**). Although the total yield could be brought up to the level of not waterlogged soil by the extra dose of N-fertilizer >105 kg N ha^{-1} (**Table 4**), the flooded areas had later maturing and up to three generations of tillers. There was no effect of other plant nutrients than N (**Table 3**).

Fertilizer	Yield kg ha^{-1}	Yield relative	Protein, %	Kg N in cereal yield, kg N ha^{-1}
None	3290	100	10.3	46
30 kg N ha^{-1} $CaNO_3$	3970	121	10.5	57
60 kg N ha^{-1} $CaNO_3$	4690	143	11.4	73
60 kg N ha^{-1} $CaNO_3$ + 28 kg K ha^{-1} KCl	4650	141	11.2	71
60 kg N ha^{-1} Yara Fullgjødsel® 21–4-10	4640	141	10.9	69
90 kg N ha^{-1} $CaNO_3$	5080	154	11.9	82
LSD, 5%	250		0.3	

Table 3.
Effects of fertilization on yield of cereals, spring barley and oats (15% water content), protein content (% of DM), and uptake of N after 2–9 days of waterlogging during the flood in 1995. Means of 14 fields in Østerdalen and Gudbrandsdalen, Norway, modified after [9].

Treatment	Extra N by farmer, kg ha^{-1}	Extra N, experiment, kg ha^{-1}	Grain yield, barley, kg ha^{-1}
Not waterlogged	0	0	6750
Waterlogged 2–3 days	0	0	3960
Waterlogged 2–3 days	0	30	4610
Waterlogged 2–3 days	0	60	5940
Waterlogged 2–3 days	0	90	6430
Waterlogged 2–3 days	45	0	5270
Waterlogged 2–3 days	45	30	6220
Waterlogged 2–3 days	45	60	6960
Waterlogged 2–3 days	45	90	7180

Table 4.
Effects of N-fertilization after flooding on yield of barley (kg ha^{-1}, 15% water content) at Selsvollene in Gudbrandsdalen, Norway.

3.2 Restoration of areas damaged by erosion

The cereal yields varied much between the different types of restoration measures at Øksna. When the measures were grouped into three: eroded topsoil or no damage (Types No. 2, 3, 5), leveled areas (Types No. 6 and 8), filled erosion craters (Types No. 1, 4, 7, 9, 10, 11), significant difference was found between the undamaged areas (eroded topsoil/no damage) and the restored areas. Average of the restoration measures gave 85% of the yields compared to the areas with eroded topsoil or intact soil (**Table 5**). There was large variation within the group filled erosion craters. Type No. 4 gave 5340 kg ha^{-1} yield of cereals as mean for the period 1999–2002 and type No. 11 gave only 2920 kg ha^{-1} cereal yield in the same period. The results indicated that it is possible to find combination of treatments that can fully restore the productivity after severe erosion caused by flooding.

The two types of sewage sludge had different effects on the yields. The limed and calcareous type had positive effect. The types Nos. 1 and 4 were filled with the similar loamy fine sand [**Table 5**], while sewage sludge was mixed in the topsoil at No 4. Although the difference was not statistically significant, the average yield for No. 4 was about 500 kg ha^{-1} higher in the period 1999–2002 than for No. 1. Most of the treatments had low levels of readily available P in the topsoil (<5 mg 100 g^{-1}), which may result in

Treatment	Type No.	Yield of cereals, kg ha^{-1}
Eroded topsoil or intact topsoil	2, 3, 5	5280a
Leveled areas	6, 8	4450b
Filled erosion craters	1, 4, 7, 9, 10 11	4520b

Table 5.
Average yield of cereals (spring barley and oats) (kg ha^{-1}, 15% water content) at plots with different restoration measures at Øksna, Elverum, for the period 1999–2002. Means followed by different letters are statistically significant ($p < 0.05$).

Type No.	Sand, 0.06–2 mm	Silt, 0.002–0.06 mm	Clay, <0.002 mm	Bulk density, Mg m^{-3}	Porosity, $m^3 m^{-3}$	TOC, g 100 g^{-1}	pH (H_2O), 1999	pH (H_2O), 2002
1	58.5	36.8	4.8	1.51	0.41	0.87	6.3	6.0
2	41.4	52.0	6.5	1.32	0.47	0.97	6.3	6.3
3	39.5	54.0	6.5	1.48	0.41	1.20	6.8	6.3
4	53.4	41.3	5.2	n.d.	n.d.	1.27	7.0	7.8
5	61.7	29.3	9.0	1.32	0.49	1.25	6.4	6.4
6	21.0	68.5	10.6	1.15	0.53	1.15	6.1	6.5
7	21.2	72.7	6.1	1.40	0.48	0.85	6.4	6.8
8	36.5	56.4	7.0	n.d.	n.d.	1.00	6.5	6.5
9	57.4	38.2	4.3	1.32	0.47	0.45	5.7	6.4
10	43.9	50.5	5.6	1.39	0.46	1.75	6.2	6.1
11	70.3	26.4	3.4	0.65	0.68	2.00	5.0	7.0

Table 6.
Soil texture, bulk density, porosity, TOC, and pH in topsoil (Ap-horizon) at Øksna for different restoration types.

P-deficiency for the crop [10]. The limed sewage sludge increased the level of readily available P in the topsoil significantly, resulting in very high levels of P-AL (>35 mg 100 g^{-1}). The other type of sewage sludge gave no additional yield compared to the other types of restoration measures. Although both sludges contained much phosphorus, soil analyses indicated deficiency of phosphorus after application of the sludge with Al salts used as coagulants. The plant availability of P in sludges produced from wastewater using Al salts as coagulants is often low compared with that in mineral fertilizer, while liming can increase the P fertilization effect of sludges [11]. This may explain the difference in effect between the two types of sewage sludge applied at Øksna.

There was no significant difference in TOC in the topsoil between areas with intact or eroded A-horizon and areas where the topsoil was amended with sludge during the restoration and the content of TOC was rather low, mainly 1 ± 0.3 g 100 g^{-1} (**Table 6**). The type No. 11 where peat material had been mixed in the topsoil had low pH initially and was limed before the growing season 2000. Although this treatment had good physical properties for plant growth, this treatment had the lowest content of available plant nutrients and different plant deficiency symptoms were observed on the cereal plants (P, K, Zn) resulting in low yield of cereals.

4. Discussion

The tolerance to waterlogging varies between different cereal crops. Cereals as barley and wheat may produce adventitious roots with about 20% of aerenchyma [12]. Photosynthesis may continue under waterlogged conditions. In the present study, waterlogging up to 5 days reduced the relative wheat yield to 60–82% of the not waterlogged reference. This is within the range of yield decrease reported in literature [13]. Flooding by cold running water may cause the soil not to be anoxic, and the yield loss could be decreased compared to situations with higher water temperature and stagnant water [14]. During the large flood in 1995, cold melt water combined with rain gave relatively low temperatures in the water. Therefore the crops could survive for some days under waterlogged conditions. In 1995, the mean day and night temperature was below 6°C until 21st May and during the flood, the maximum day and night temperature was 15–18°C in the lowland parts and 8–10°C in the mountain area [4].

As the cereal yields could be brought up to the level of areas without waterlogging by sufficient N-supply after the flood, it seems reasonable that the period of flooding had delayed the growth due to the cold water. Nitrogen had most likely been lost by leaching during the flood, while denitrification may have occurred at sites with stagnant water.

The flood in July 1789 occurred as a result of 3 days of continuous rain combined with snowmelt of snow and ice in the mountainous areas. This led to a combination of landslides from the hillslopes and enormous flood discharge in the large rivers Gudbrandsdalslågen and Glomma and also in smaller side rivers [15]. Prior to this catastrophic flood, there had not been any large snow melt flood for many years and very cold winters. This can be related to the eruption of the volcano Laki at Iceland in 1783–1784. The effects of the aerosols and extreme volcanic pollution (i.e., dry fog) that effected Europe and other regions in 1783 have been estimated to cause a drop in temperature of 1.3°C in Europe and North America, lasting for 2–3 years [16]. After the flood in July 1789, large amounts of sand (50–100 cm thick) had sedimented above fertile soil with sandy loam texture in several places in Gudbrandsdalen. These areas gave very low yields and were subjected to drought. A deep plowing project showed promising results when the old plow layer was brought back to the soil surface. However, deep plowing gave only stripes with good effect [17], while use of excavator for bringing the old fertile soil to the surface fully restored the productivity of the soil. ^{14}C dating of buried peat material below the sediments from the 1789 flood gave interesting results. In the peat layers it was found layers and lamina of silt up to 6–7 cm thickness between the organic layers. The difference in age between the bottom peat layer and the layer just below the thick mineral sediments was 4500–5000 years [5]. This indicates that the 1789 flood was an exceptional event, which had not happened for several millennia.

The reconstructed soils at Øksna and other areas damaged by erosion and sedimentation by the 1995 flood have been protected by building higher levees and similar measures, which prevented damage by a smaller flood in 2013. When agricultural lands close to large rivers are subjected to flooding in spring almost every year, the flood situation will have large impact on the ground water table in the soils at fluvial plain. Periods with high groundwater level may have large impact on the yields both on cultivated Fluvisols and reconstructed soils along the rivers, as found in a study in Nedre Eiker along the River Drammenselva, north of the city of Drammen, Norway [18].

Outburst floods from glacial lakes about 10,000 years BP have significantly influenced the geomorphology of the main valleys in eastern Norway, Østerdalen and Gudbrandsdalen. Silty sediments of 0.5–1 m depth above marine clay deposits at Romerike, more than 200 km south of the start point of the outburst flood indicate flood of enormous dimensions [6, 19]. Similar to the soils at Øksna, the silty soils at Romerike were poor in nutrient content and cultivation of these silty soils took place in the 1920s and 1930s after trace elements were included for fertilization and appropriate drainage techniques were applied [20]. Documentation of such flood sediments is of importance for prediction of effects of flood caused by dam failure of a hydroelectric plant.

5. Conclusion

The investigations after the flood in central Norway in 1995 showed that flooding of 2–9 days duration caused decreased yields of cereals as spring barley, spring wheat, and oats. As sufficient N-supply after end of the waterlogged period could bring the yields of barley and oats up to normal levels, it is likely that the flooding with cold running water caused delayed development and maturing of the crop and the fertilization could compensate for N leaching during the flood.

Reconstruction of erosion craters and other measures to make the damaged agricultural areas into agricultural production showed large differences between the best and the least successful measures. The restoration measures gave as a mean about 85% of the yield compared with areas without damage or eroded topsoil. Filling erosion craters with loamy fine sand and mixing in limed sewage sludge gave similar yields as the undamaged areas. This treatment significantly increased the amount of readily available P in the soil, while use of another type of sewage sludge had no positive effect on the yields.

Acknowledgements

The investigations after the flood in 1995 were supported by the Norwegian Research Council, Grant No. 110038/110. The study of effects of measures for reconstruction of soils after the flood in the period 1999–2002 was supported Statens landbruksforvaltning, (now named Landbruksdirektoratet). Harvesting of the plots was performed by Norsk landbruksrådgiving, Innlandet. The farmers Gunnar Sætersmoen and Helge Øverseth at Øksna are thanked for good cooperation.

Author details

Trond Knapp Haraldsen
NIBIO, Norwegian Institute of Bioeconomy Research, Division of Environment and Natural Resources, Ås (Aas), Norway

*Address all correspondence to: trond.haraldsen@nibio

References

[1] Paprotny D, Sebastian A, Morales-Nápoles O, Jonkman S. Trends in flood losses in Europe over the past 150 years. Nature Communications. 2018;**9**:1985. DOI: 10.1038/s41467-018-04253-1

[2] IUSS Working Group WRB. World Reference Base for Soil Resources 2014, update 2015. International soil classification system for naming soils and creating legends for soil maps. World Soil Resources Reports; 2015. No. 106. FAO, Rome

[3] Brémond P, Grelot F, Agenais A-L. Review Article: Economic evaluation of flood damage to agriculture – Review and analysis of existing methods. Natural Hazards Earth and System Sciences. 2013;**13**:2493-2512

[4] Øygarden L, Eggestad HO, Standring WJF, Goffeng G, Vagstad N, Flommen i. Skader på jordbruksarealene langs Glomma og Gudbrandsdalslågen. Jordforsk Report. 1995;**29**(1996):41

[5] Haraldsen TK. Stor-Ofsen – et 200 års minne. Jord og myr. 1989;**13**(3):89-101

[6] Longva O. Flood Deposits and Erosional Features from the Catastrophic Drainage of Preboreal Glacial Lake Nedre Glåmsjø. SE Norway: Department of Geology, University of Bergen; 1994

[7] NIBIO Kilden. Available from: https://kilden.nibio.no [Accessed: 2022-09-04]

[8] Egnér H, Riehm H, Domingo WR. Untersuchungen über die chemische Bodenanalyse als Grundlage für die Beurteilung des Nährstoffzustandes der Böden. Kungl Lantbrukshögskolans Annaler. 1960;**26**:199-215

[9] Stabbetorp H, Hiitola K. Undersøkelser of forsøk i forbindelse med flommen i 1995. Jord og plantekultur 1996, Planteforsk. 1996. pp. 70-74

[10] Krogstad T, Øgaard AF, Kristoffersen AØ. New P recommendations for grass and cereals in Norwegian agriculture. In: Rubæk G, editor. Phosphorus Management in Nordic-Baltic Agriculture – Reconciling Productivity and Environmental Protection. Uppsala, Sweden: Nordic Association of Agricultural Scientists (NJF); 2008. Report 4(4). ISSN 1653-215

[11] Øgaard AF, Brod E. Efficient phosphorus cycling in food production: Predicting the phosphorus fertilization effect of sludge from chemical wastewater treatment. Journal of Agricultural and Food Chemistry. 2016;**64**:4821-4829. DOI: 10.1021/acs.jafc.5b05974

[12] Ploschuk RA, Miralles JD, Colmer TD, Ploschuk EL, Striker GG. Waterlogging of winter crops at early and late stages: Impacts on leaf physiology, growth and yield. Frontiers in Plant Science, Section Crop and Product Physiology. 2018;**9**:1863. DOI: 10.3389/fpls.2018.01863

[13] Herzog M, Striker GG, Colmer TD, Pedersen O. Mechanisms of waterlogging tolerance in wheat–A review of root and shoot physiology. Plant Cell and Environment. 2016;**39**:1068-1086. DOI: 10.1111/pce.12676

[14] Setter TL, Waters I. Review of prospects for germplasm improvement for waterlogging tolerance in wheat, barley and oats. Plant and Soil. 2003;**253**:1-34. DOI: 10.1023/A:1024573305997

[15] Sommerfeldt W. Ofsen i 1789 og virkninger av den i Fron. Fron Historielag; 1972. Reprinted version of

master thesis in Geography, University of Oslo, Norway, originally published in 1943. p. 236

[16] Thordarson T, Self S. Atmospheric and envriromental effects of the 1783-1784 Laki eruption: A review and reassessment. Journal of Geophysical Research-Atmospheres. 2003;**108**(D1):4011

[17] Njøs A. Pløying til stor dybde. Djuparbeiding av lagdelt sandjord i Sel i Gudbrandsdalen. En vurdering av pløyevirkningen ut fra kjemiske- og fysiske jordanalyser. Ny Jord. 1964;**51**(2):47-57

[18] Anda TN. Transportation and construction of agricultural soil as a method for sustaining production in agricultural areas after anthropogenic encroachment – A survey of transported agricultural soil in Nedre Eiker. [MSc. thesis]. Norwegian University of Life Sciences, Ås, Norway. Available from: https://nmbu.brage.unit.no/nmbu-xmlui/bitstream/handle/11250/2398784/Anda.2016.pdf?sequence=1&isAllowed=y [Accessed: 2022-09-04]

[19] Rutherford LA. The properties, distribution and origin of white silt soils at Romerike, Norway. I. Profile morphology, geographic distribution and pedological properties. Meddelelser fra det norske skogforsøksvesen. 1972;**30**:139-168

[20] Sveistrup TE, Haraldsen TK, Langohr R, Marcelino V, Kværner J. Impact of land use and seasonal freezing on morphological and physical properties of silty Norwegian soils. Soil & Tillage Research. 2005;**81**:39-56

Chapter 6

Post-Fire Debris Flow Susceptibility Assessment Tracking the "Cauliflower Effect": A Case Study in Montecito, USA

Johnny Douvinet

Abstract

Most of the studies focused on triggering conditions to identify the critical thresholds beyond which the occurrence of postfire debris flows becomes more than likely. However, researchers rarely focused on the relations between the morphological patterns and influences on surface water flows, while after extreme fires, the burned areas strongly reduce the infiltration capacities and generate important runoffs. So, to address these relations, we used the cellular automaton RuiCells©. This model brings out the concentration areas inside a given form, in which networks and surfaces are well-structured, and patterns are similar to efficient forms that can be found by looking at a cauliflower. This model has been applied to assess the flash floods susceptibility in sedimentary areas, with a success rate of 43%, so we decided to apply this model to the five catchments located at the apex of urbanized fans upstream of Montecito (Santa Barbara County, USA), affected by debris flows that occurred on January 9, 2018, 20 days after the Thomas Fire (one of the largest wildfires in California history). Some of values have never been observed elsewhere. So, we might apply this approach to assess the postfire debris flows susceptibilities given the increasing number of fires and mega fires.

Keywords: debris-flows, postfire conditions, spatial behavior, Montecito, cauliflower effect

1. Introduction

Given projected increases in wildfire size and severity [1], precipitation intensity and variability [2, 3], or in development in the wildland-urban interface [4], there is a growing need to increase resilience to disasters [5] and to reduce as far as possible impacts of hazards on lives, properties, bridges, roads, and infrastructures [6]. Postfire debris flows are phenomena able to rapidly transport large volumes of sediment and large boulders, sometimes over long distances, making surface flows destructive and dangerous [7–9]. The debris flow dynamics are determined by solid and fluid forces, while in floods and hyper-concentrated floods, the dominant process is more determined by fluid forces alone [10]. The transition between processes represents a spatial and temporal continuum: one single event is often related to different pulses that have different characteristics [11]. Furthermore, the flow properties vary along the course of the channel with the lower channel reaches often exhibiting flood characteristics due to increased sediment deposition rates as well as dilution from increased water content relative to the sediment entrainment rate [12, 13].

Currently, evidence has emerged on postfire rainfall thresholds and the relations between convergence zones and preexisting drainage lines [14], which results in rapid channel development where bedrock eventually set the lower limit of scour depth. Immediately after a fire, the role of overland flow during rains becomes magnified due to losses of vegetation [15, 16], changes in soil properties, and sediment supply [17]. Thresholds are significantly lower than most identified for unburned settings, due to the difference between rapid runoff-dominated processes acting in burned areas, and longer-term, infiltration-dominated processes on unburned hillslopes [9]. However, the hydro-geomorphic response of burned upland regions can be variable. It depends on various factors including the fire severity, timing, and properties of postfire rainfall events, as the inherent geomorphic and hydrological characteristics of fire-affected catchments [1, 3]. Debris flows or sediment-laden floods are produced from the small burned catchments, in response to short rains and convective thunderstorms in the intermountain west U.S. [18–20], and to longer duration winter frontal storms in southern California [21, 22]. Therefore, unlike landslide-triggered debris flows, these events have no identifiable source, and they can occur with little or without moisture.

However, researches rarely focused only on the links between the morphological patterns and influences on surface water flows, while after extreme fires, the burned areas strongly reduce the infiltration capacities and generate high Hortonian runoffs. Indeed, the debris flow susceptibility still remains assessed by considering the slope, curvature, elevation, or terrain complexity as secondary factors [23, 24]; meanwhile, the impacts of network and surfaces, organized within a given form, are neglected, whereas the morphological effects play a strong influence during postfire conditions.

To overcome such a problem, this study proposes to apply a methodology that we have already tested in the situation of high Hortonian runoffs [25–27]. The cellular automaton (so-called RuiCells©) is used to track efficient points upstream of which networks and areas are well-structured, and patterns are similar to efficient forms that can be found by looking at a cauliflower. This model has been applied in France to assess the flash flood susceptibility in sedimentary context, with a success rate of 43%, so we decided to apply this model in other areas or for more extreme events such as postfire debris flows. If results are positive, this model could then bring a new way to assess the postfire debris flows susceptibilities in more catchments.

After presenting the five studied catchments and the 2018 debris flow features (Section 2), the method and data used are described (and especially the "cauliflower effect") in Section 3. Results are described by following two observation levels: the morphological signatures obtained at the global scale and the "cauliflower effect" detected at fine scales (Section 4). The discussion continues by focusing on the relations between the importance of burned areas and usefulness of debris basins, in relation to the "cauliflower effect" and morphological influences (Section 6).

2. Hydrological data and studied catchments

2.1 Study area

Located 8 km east of Santa Barbara, Santa Barbara County, California (USA) (**Figure 1**), Montecito is an unincorporated community and a census-designated place. The population was estimated in 2018 to be 9145 residents, and Montecito is regularly ranked by the Forbes magazine as one of the wealthiest in the United States (2016, 2017). According to the magazine, it is 21 of the 100 most influential public figures in the United States are known to own property (2017). The climate is characterized by warmer winters and cooler summers, compared with places further inland, because of the ocean's proximity [5, 28]. Located at the foot of the Santa Ynez Mountains, which are mostly of sedimentary origin, the peak relief has an altitude of 981 m. Several creeks span approximately 3 km between the mountain front and the Pacific Ocean, intersecting State Route 192 and Highway 101. Since distances between mountains and bay are shorter, steep terrain presents hillslope gradients: ~37% of the terrain exceeds 35 degrees, and creek-bed gradients are ~12% [5]. Under these conditions, surface runoff and waters rapidly flowed south into a series of creeks with gradients of ~5%, and drain south through the residential city of Montecito, with construction in a series of alluvial fans [29, 30]. The urbanized piedmont plain contains steeply sloping alluvial-fan landforms generally north of State Road 192 and Highway 101 and gently sloping alluvial-fan landforms near the coast. And to protect the high density of roads and structures on the alluvial fans, four sediment-retention basins have been built along the main paths: Cold Spring (1964), San Ysidro (1964), and Romero (1971), as recorded by Santa Barbara County.

2.2 Awareness of the risk of debris-flows

Before the 2018 event, Montecito had a high level of situational awareness prior to the storm [5]. Historical and damaging debris flows (1926, 1934, 1964, 1969, 1971, 1990, and 2002) have been already registered [31, 32], and debris flows became a

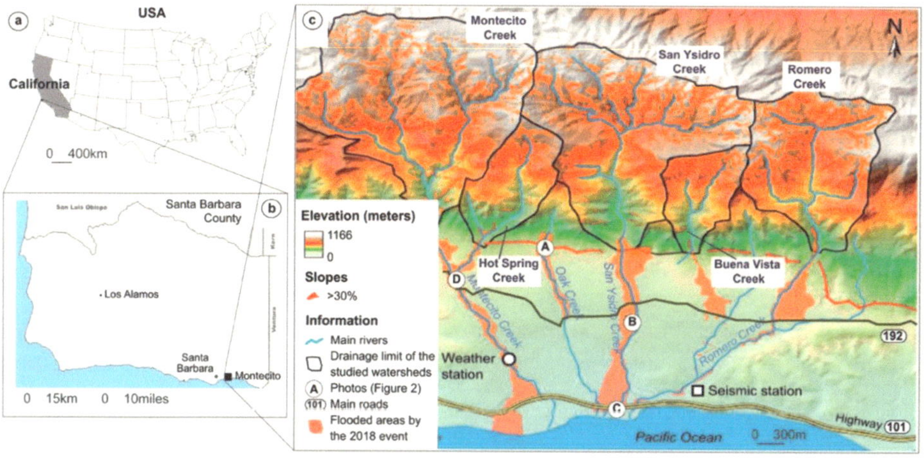

Figure 1.
Location of the five studied catchments (with damage extracted from 31).

Catchment Name	Catchment area (km^2) – *A*	Basin slope (50th, 90th percentiles)	Area burned at moderate to high severity (%)	Predicted likelihood for design storm. (I_{15} = 24 mm/hr)
Montecito Creek	14.07	28/39	79	65%
Hot Spring Creek	3.34	28/40	49	70%
San Ysidro Creek	11.63	28/40	85	69%
Buena Vista Creek	2.66	29/41	82	63%
Romero Creek	7.11	30/42	78	62%

Table 1.
Indicators confirming the high probability for a postfire debris flow in Montecito just before the 9 January 2018.

topic of research since the 1934 debris flow in Montrose that killed over 40 people [29, 30]. Some researchers documented the sediment-laden water flow following the 1990 Painted Cave fire, and they have proven that the first few months following wildfire are of most concern [29, 30]. Awareness of the debris flow risk was emphasized by coordinated efforts between county, state, and federal agencies that included: (1) a determination of the soil burn severity as a good indicator of flooding and debris-flow potential [33]; (2) the debris-flow hazard assessment that shows the high likelihood and potential volume of debris flows from the burned areas, in response to design storms [34]; a warning system that predicted significant to the extreme potential for debris flow in the 4 days leading up to the storm [35, 36]; and (4) a proactive emergency community that coordinated evacuation orders to reduced casualties [5]. However, despite these efforts, it remains hard to anticipate such events: (1) assuming the annual probability of a wildfire in this area (following recovery from a previous fire rated at 0.10), and that the probability of a short duration high intensity storm is 0.02, the conditional probability is 0.002 [29, 30]; (2) prior to the 9 January 2018 rain event, the U.S. Geological Survey (2018) had predicted a 62–70% probability of debris flows from the five catchments located in upstream Montecito (**Table 1**) for a design rainstorm of I15 = 24 mm/hr. [34], but forecasting such intensity remains difficult in real time.

2.3 Characteristics of the January 2018 event and hydrological data

The Thomas fire above Montecito burned from the ridge crest of the Santa Ynez Mountains on December 2017 (from the 14th until the 27th), to approximately the apex of the urbanized alluvial fans [5, 6, 29, 30]. Under these conditions, the occurrence of debris flows was feared, and rainfalls were watched [37]. Early on the morning of 9 January 2018, a heavy rain characterized by high intensity (13 mm in 5 min, e.g. 157 mm/h) at 3.45 a.m. (recorded by the Montecito rain gauge) caused flows consisting of mud, boulders, and tree branches up to 15 feet (up to 5 m) in height, and moving at estimated speeds of up to 20 miles per hour (around 30 km/h) into the downstream creeks, valleys, and lower areas of Montecito [38]. A few minutes later, at 3:47 a.m., the Montecito Fire Department received the first calls of a major explosion on San Ysidro Creek–East Mountain Drive [5]. Flows on Cold Spring Creek began at 3:49 a.m. (this was confirmed by the security camera footage). Inundations began on the northern plains of Montecito around 3:50 a.m., and estimated lag times are estimated ranging from 4:06 to 4:10 a.m. [39], with the seismic signature of debris flows. The rainfall threshold for the occurrence of debris flows decreased obviously due to the Thomas Fire and the percent of burned areas, ranging from 49% within Hit

Catchment Name and number	Catchment area (km^2) – *A*	Accumulation Area (km^2) – AA	Inundated Area (km^2) – IA	Estimated Sediment Volume (m^3) - OSV
Montecito Creek	14.07	0.78	0.997	231,000
Hot Spring Creek	3.34	0.08	0.102	10,000
San Ysidro Creek	11.63	0.69	0.905	297,000
Buena Vista Creek	2.66	0.11	0.290	41,000
Romero Creek	7.11	0.17	0.312	100,000

Table 2.
Data collected during the 2018 Montecito event [5, 6, 34, 37].

Spring Creek to 85% in San Ysidro Creek (**Table 2**). No floods occurred on 9 January 2017 (for cumulative precipitation of 16.5 mm), whereas several debris flows occurred on 9 January 2018, with cumulative precipitation of less than 13.7 mm in 5 min [28]. And the sediment basins do not limit the violence of flows.

The 30-foot wall of boulders, mud, and debris flows ran down hillsides at 15 miles per hour, injuring dozens and causing 21 prehospital deaths and 2 missing persons (two children of 17- and 2-year old's). 163 people were hospitalized for injuries [40] and a retroactive review conducted of 24 victims from the debris flows presenting to Cottage Health showed that most part of symptoms referred to soft tissue injuries (100%), hypothermia (67%), craniofacial injuries (67%), corneal abrasions (53%), and orthopedic injuries (47%), as well as the loss of an immediate family member (73%) because of the incident [40]. Around half of the victims who perished were migrants from working-class families [28]. Hopefully, crews rescued at least 50 people by air, dozens more from the ground, and 300 people stocked in their homes along the Romero Canyon neighborhood after impassable roads halted rescue operations, the disaster caused $177 million of insured property damage, $7 million in emergency responses, and $43 million in cleaning costs [41, 42], with a final cost approaching more than $250 million in 2019 [5].

Although creeks are incised by more than 5 m into the surrounding terrain (**Figure 2**), the debris flows overflowed the valleys, often at bridge crossings, and carried boulders into the neighboring residential areas [5]. The debris flow deposits cover around 7 km^2 and the cumulative amount of sediment are ranging from 297,000 m^3 [33], 680,000 m^3 (including boulders up to 6 m. in diameter) to up to 880,000 m^3 [34]. Damage was concentrated within the 500-meter-wide flow path in numerous areas, and was mostly pronounced along the Montecito and San Ysidro Creeks, as 80% of the 408 damaged and destroyed homes were located in them. Additionally, flow bifurcated approximately 0.7 km downstream of the mountain front and traveled in a southwest direction from San Ysidro Creek, along El Bosque Road [37]. Finally, debris flows resulted in damage to at least 163 structures and complete destruction of an additional 92 structures.

3. Methodological assessment

To complete researches that have been already carried out on the 2018 Montecito event [5, 35–39], the deterministic cellular automaton so-called RuiCells© was used.

Figure 2.
Damage and spatial extent of debris flows in 2018 in Montecito.

First, to further assess the dynamical influences of morphological conditions on the sudden hydrological responses of the catchments located upstream of alluvial fans. Any morphometric parameters have been reported to be related to debris flow potential [25, 26], but their usefulness is limited and often criticized [27]. Second, to track the points in upstream of which areas and networks are well organized (the so-called "cauliflower effect"). Simulations give a picture of the "*width function area*" [25, 27], but concentration area or internal efficiency can be detected, and this effect could explain violent hydrological responses, in time and space.

3.1 The "cauliflower effect" and its links with hydrological features

The series of processes from the cause generating a cauliflower form is complex. One might assume that the underlying rules for the cauliflower growth are simple, even if the form is of great complexity [43]. But the crucial phenomenon that ultimately leads to much of the structure is that at the tip of a stem is possible for new stems, to form and branch off. In the simple cases, these new stems are in essence just smaller copies of the original stem. With this setup, the branching succession can be represented by steps in the evolution of a neighbor-independent substitution system. The cauliflower finally presents an unusual phyllotaxis, with a multitude of spirals nested over a wide range of scales. This self-similar organization culminates in the Romanesco cultivar, where the spirals appear in relief due to their conical shape at all scales, a geometrical feature conferring the whole curd a marked fractal-like aspect [43].

Similarities can be found by comparing forms of cauliflower and hierarchical river organization: without geological or lithological constraints, a stream river branching progress through scales [44], and if the distance before the stream appears is determined by the rates of production or the erosive capacity of waters, a minimum area is needed for elementary catchments [45]. The layout of the network becomes fundamental considering the path and transit time, from the source to the outlet. The *width function* created by [46] has permitted consideration of the number of hydrological links located at equal distances from the outlet [46], thus taking into account the network and the surfaces within a given watershed form. This function remains today one of the most relevant tools, to link the shape of a basin and its hydrographic network, to the hydrological response resulting from this organization [25, 27], even if numerous works have improved the calculation and the method of extraction of this function, by associating the slopes. A structurally well-organized network finally presents, without external constraints, a form nearest to those of a cauliflower: it represents a homothetic phenomenon, with a minimal dispersion of energy, which reminds the "self-organized critical systems" [47]. But the global catchment form can be hidden on the global scale. Then, we create a specific model to measure this effect.

3.2 Simulation of the "cauliflower effect" with the RuiCells© model

To address the "cauliflower effect," we used a specific cellular automaton so-called RuiCells, that has been already described before [26, 27]. For this specific simulation, we simulate the sum of surfaces flowing within networks. Cells at the beginning of the simulations are initialized with their surface (defined by the TIN and depending on the DEM resolution, here 50 m as previous studies have been calibrated on this). RuiCells© handles the advection operator in moving surfaces between each cell, so a formal property defined in classical CA is maintained [48], as transition rules operate on cells based on local neighborhood. Surface flow is routed via each cell until the downstream boundary is reached [48]. At the end of the simulation, a graph presents sum of surfaces registered at each interaction to present the morphological signature of the basins, defined as a function of distance *n* from the outlet (**Figure 3**). Steps are length steps since the surface flow diffusion depends on the spatial lattice size. These surface flow graphs give the 2D theoretical spatial behavior of a catchment. This improves older methods, that only informed on the number of links in the network at a flow distance x to the outlet [46], as areas flowing in RuiCells© follow hydrological rules (differences between surface, linear, or node transition are accounted), and are based on a triangular lattice.

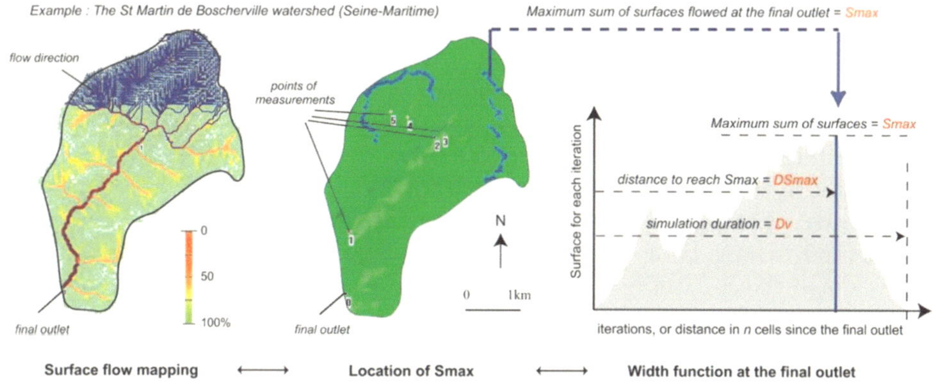

Figure 3.
Sum of surfaces per iteration, simulated by RuiCells©.

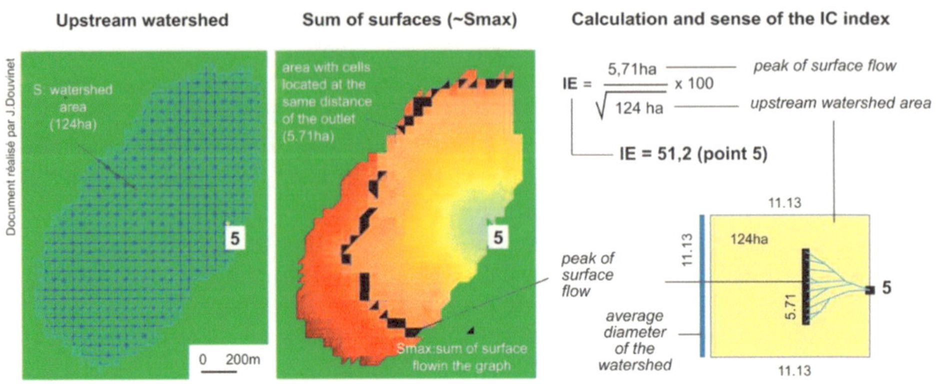

Figure 4.
The concentration index (CI) estimated by the RuiCells© model.

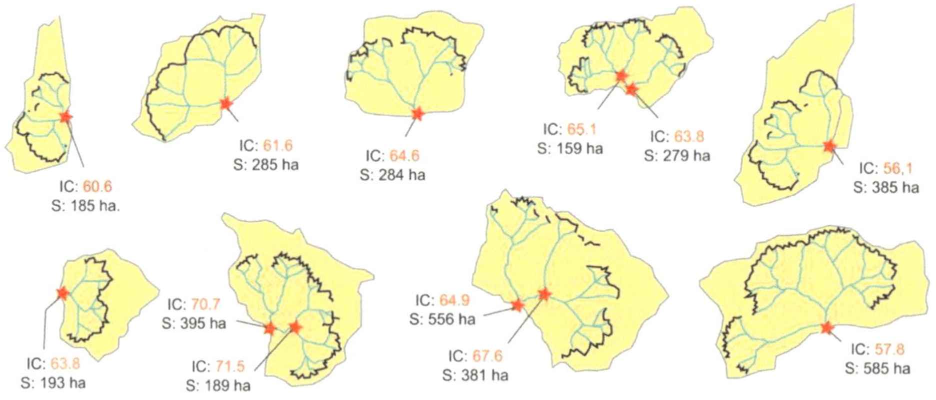

Figure 5.
Line of SMax in catchments where runoff concentrate in a short distance (according to Douvinet, 2008): the so-called "cauliflower effect."

In addition, to track the "cauliflower effect," the highest sum of surfaces (*Smax*) is divided by the square root of the catchment area (A') located upstream (**Figure 3**), to define a Concentration Index (CI). *Smax* equals to the highest line of cells located at a distance n from the point of measurement. As *Smax* is related to a length (since the square root equals to the average diameter of the upstream area), CI is dimensionless (**Figure 4**). Values are automatically calculated during the simulation process and are mapped. Strong values (colored in red) identify the neuralgic points inside the basin where the network is very effective with respect to the shape in which it is inserted. Values have the same significance regardless of the basin area if the TIN resolution never changes. In this approach, we always use a DEM of 50-m long. If CI equals to 50, it indicates a medium flow concentration: the peak of surfaces (*Smax*) corresponds to half of the average diameter. But if CI values exceed 55, networks and surfaces appear highly well-structured [25]. If strong values are observed, the efficient line of cells *Smax* is mapped, and a perfect form (so-called the "cauliflower effect") appears. Examples (**Figure 5**) illustrate this in previously studied catchments: red stars located the efficient points, blue lines some of the branches of the networks, and black lines the surfaces. Points and linear with IC > 55 concentrate high surface flows in a short distance, with a minimum energy expenditure [27, 47].

3.3 Prospects and limitations

Simulating the influence of morphological conditions on hydrological responses is limited for a few reasons: (1) this theoretical approach cannot be compared with real simulation results obtained with more pervasive models; (2) the efficient points can be located within catchments where no violent floods occurred (probably due to the lack of intense rains); (3) sediment production is not combined with water production, while these processes could complexify the real hydrological response.

Estimating the "cauliflower effect" in the five catchments located at the apex of the Montecito may, however, allow: (1) to check if the erosive areas are related to sudden debris flows and important damage; (2) to compare information supported by the "cauliflower" effect to other results. If the model identifies real sensitive areas, it could extend the model's usefulness and the scope of the model. If the model identifies sensitive areas that have not experienced severe flooding, the rainfall can have not been sufficient to cause them to respond, and the lack of correlation could raise concerns about a future event. If the model does not identify sensitivity in areas that have been affected, the relevance of this model may be excluded, and the morphometric factors should be excluded as efficient, to look for the origin of the susceptibility in the other variables (soil, burned areas, rain, exposure...). This validation process finally allows us to know if RuiCells© can be useful for the decision-makers in the era of extremes, to whom we could say: then, "if you observe a cauliflower effect in a burned catchment, be very careful in case of rains!"

4. Results

4.1 Morphological signatures and role of internal organization

The spatial behaviors simulated at the outlet of the five basins are first studied. Several points of measurements have been added within the biggest basins, to further understand the genesis of the simulated sum of surfaces. In the Montecito Creek basin (**Figure 6**), the surfaces located upstream the points 2 and 3 support together the *SMax* measured at the final outlet (**Table 3**), around the 100th iteration. This catchment presents a high IC value at the final outlet (55.74), due to the hierarchical organization of networks. Such efficiency is more effective in the San Ysidro Creek (**Table 3**): many surfaces flow since the upstream part of the point 7, and numerous surfaces are drained thanks to the well-structured networks. In Hot Spring or Buena Vista Creek, the cascading surface flow system is slightly less inefficient as the contributions of sub-basins are shifted in space. As a consequence, values for *Smax* are lower. In Romero Creek, upstream areas of the point 10 present an efficient organization, but their contribution is not combined with other sub-basins at the final outlet, explaining the small value for *SMax* at the final outlet (**Figure 6**). A similar discrepancy induces in Hot Spring Creek a long out-flow. Obviously, the *SMax* value in upstream point 1 in the Montecito Creek is close to those obtained in upstream points 10 and 3 (**Figure 6**), but as they are not spatially combined, no major peak of surfaces can appear.

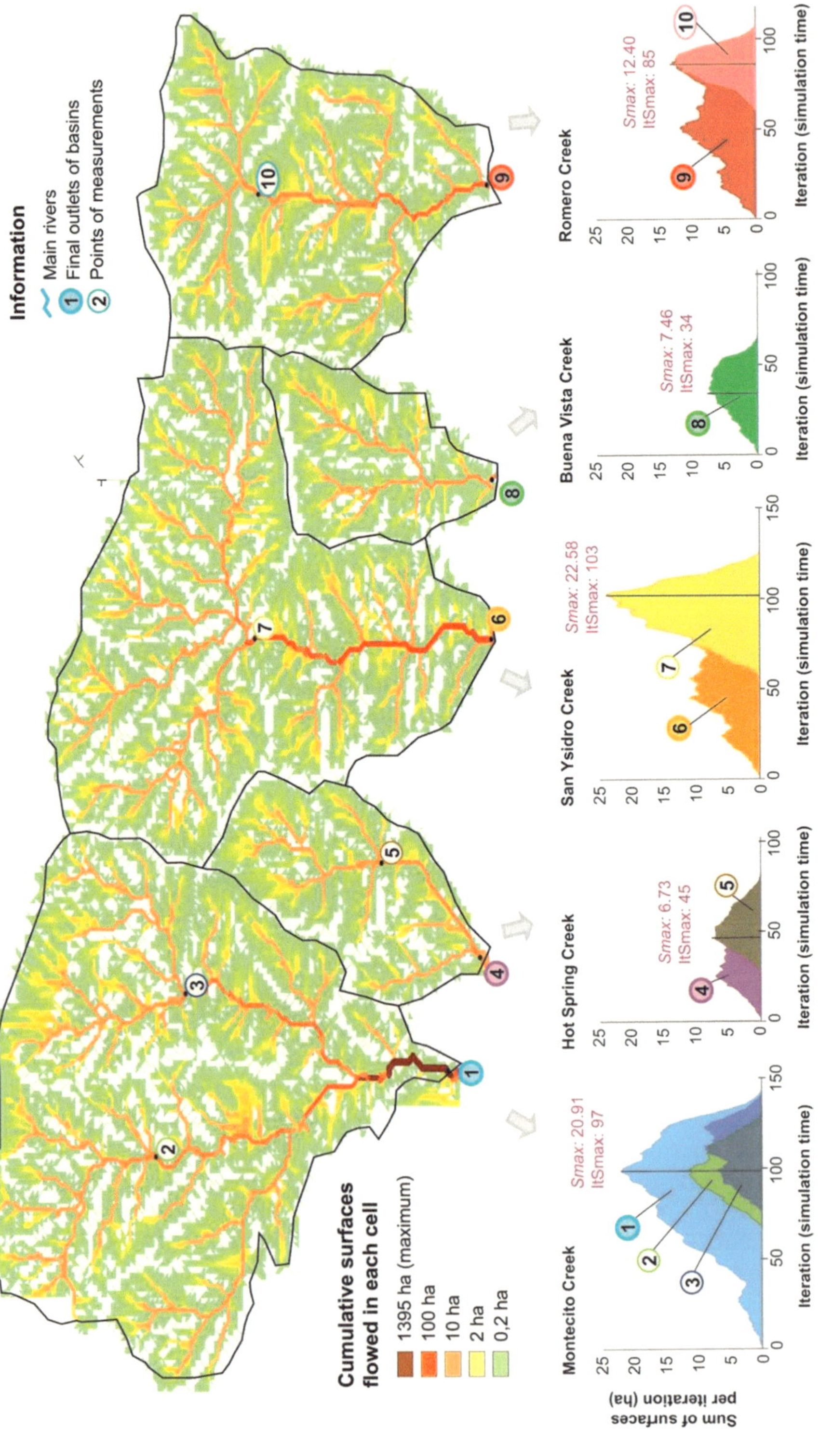

Figure 6.
Simulated surface flow response in the five studied catchments.

Catchment Name and number	Catchment area (km^2) (*A*)	Time for the simulation (It)	Maximum sum of surfaces (*Smax*)	Iteration of Smax (It*Smax*)	CI value (outlet)
Montecito Creek	14.07	147	20.91 ha	97	55.74
Hot Spring Creek	3.34	89	6.73 ha	45	36.80
San Ysidro Creek	11.63	133	22.58 ha	103	66.21
Buena Vista Creek	2.66	82	7.46 ha	34	45.74
Romero Creek	7.11	112	12.40 ha	85	46.50

Table 3.
Outputs collected at the final outlet of the five studied catchments.

This first analysis then confirms that the global catchment scale is not relevant to address the effects of morphological conditions on hydrological responses, especially for the five studied catchments: for example, the upstream part of point 7 explains 100% of *Smax* estimated at the final outlet (point 6), while its surfaces only represent 64% of the global basin size (11.63 km^2). So, we need to track concentration within networks at a finer scale, hence the interest in going down to the cellular scale.

4.2 The "cauliflower effects"

Maps indicating values of Smax at the cellular scale confirm that strong high values (IC >55) exist within the catchments (**Figure 7**). A very important value (IC = 81.1) is simulated within the San Ysidro upstream part (Smax: 22.58 ha; A: 7.74 km2), and this record has never been observed elsewhere and in previous studies. Indeed, even if RuiCells © has been applied on more than 450 catchments in France [49], the older maximum value was estimated in Saint-Martin-de-Boscherville (in France), with IC equal 71.6. In this study, for San Ysidro Creek, the internal efficiency was already suggested (**Figure 6**), but the "cauliflower effect" is remarkable: the contribution of three well-structured sub-basins suddenly increases the IC values, as they contribute together and surfaces arrive at the same moment in upstream of the point 7. IC values remain higher (>55) until the final outlet is reached (**Figure 8**), which indicates that surface flow is efficient during a distance of around 1.125 km. The number of branches and their similar distance to the outlet aggravate hydrological responses and support current solid debris content, especially during the postfire conditions.

In Romero Creek, another IC high value is clearly detected (IC = 76.9; Smax: 12.40; A: 2.65 km2), while IC was weak at the outlet (IC = 46.50). Here, the rest of the catchment does not play a role in the surface response. In Montecito Creek, several IC values appear, and they exceed the threshold value of 55 (57.8 at the point 2 and 58.4 at the point 3). One homothetic behavior is observed: efficient concentration areas emerge in different confluences in the river system, and this explains why a distance of 450 m (red-colored) upstream of point 0 still has a morphological efficiency. And finally, morphology in the Buena Vista and the Hot Spring Creeks seem to be less effective. In fact, the two basins present an internal concentration, but values are weaker (53.1 and 52.1) in comparison with others, so their morphological efficiencies are hidden by other extreme values.

As a consequence, the "cauliflower effect" is detected, and it gives new patterns to the relations between networks and areas (**Figure 8**), completing the previously

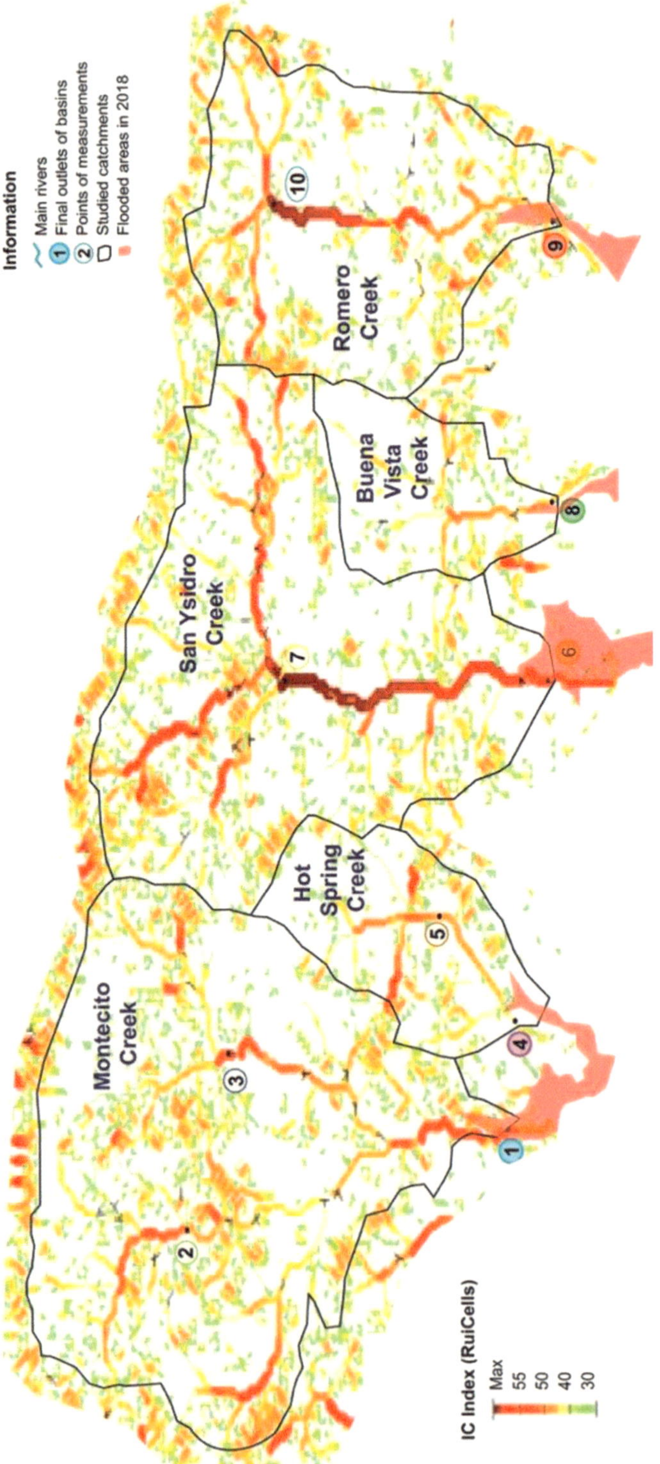

Figure 7.
Maps of IC index in the five studied hydrological catchments.

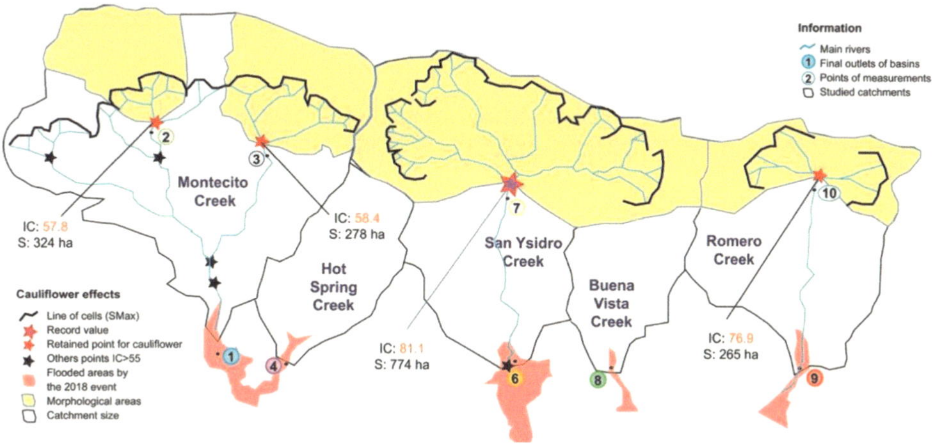

Figure 8.
The "cauliflower effect" in upstream morphological areas.

known forms (**Figure 5**). More networks and surfaces are numerous and equidistant from the outlet point, more this "cauliflower effect" takes sense as shown by the form around the new record point. The morphological areas often do not correspond to the global catchment scale, that is why the global scale is not relevant to address the morphological influence on surface flow dynamics. And it also explains with morphometric parameters (calculated for a given form or size) are really insufficient to detect their influence in case of violent and sudden events like debris flows.

5. Discussions

5.1 Insights related to previous flash floods collected data

Additionally, we compared simulation results with various data collected in other studies. Human damage was really dramatic at the final outlet of Montecito and San Ysidro Creeks: along the 1st 1 km of these two catchments from the apex, debris-flow stages were several meters above the channel banks, and at some locations exceeded 10 m above the initial thalweg [5]. By the way, our simulations demonstrate the good performance of RuiCells© model, which exhibits high morphological efficiency and the role played by morphological signature. Due to the intense efficiencies and the burned surface, conducive to hortonian runoff, debris flows arrived suddenly and violently at the interface between the final outlets and first homes located at the apex. The high burned areas (respectively, 79 and 85%) added to morphological efficiencies have also aggravated the quick responses (**Table 1**). Complementary simulations with RuiCells© [49] also proved that the rainfall intensity played a tiny role in the increase in runoff flows and volumes (1.4 to 1.86, if comparing 2017 and 2018 rainfall with no fire impacts). By injecting the 2018 rainfall and the magnitude of the burned areas (from 83 to 79% in San Ysidro, Montecito, and Romero), we estimate that fire and morphology combined with rainfall increase peak flows by orders of magnitude ranging from 9.7 to 10.3. On the other hand, other geomorphological indicators also confirm that the "cauliflowers effects" are not fictive. Large boulders (with a-axis > 1 m) were transported nearly the entire length of San Ysidro, Montecito, and

Romero creeks. But numerous locations of boulder deposits coincide with human infrastructure, roads, clogged culverts, bridge underpasses, or variations in the channel slopes. But surely all these damages are related to high water velocities and capacities, so indirectly to high concentration. And finally, topography really seems to be the key factor to assess postfire debris flows [29, 30, 34, 40].

5.2 Challenges face in the current era of extremes

The outcome of the 2018 postfire debris flow event that took place in Montecito was devastating but could have been worse if no coordination had taken place between the local, State, and Federal Agencies and if no early warning had been issued by the NWS or Santa Barbara OEM. However, our new simulation results can be used to produce a simple heuristic approach and be relevant to the existing "Duck, Cover, and Hold" for earthquakes. We might use "wildfire – intense rain – move uphill," or "Ready" (be prepared in advance to evacuate if necessary), "Set" (monitor fire burned areas and postfire precipitation in preparation to evacuate), and "Go" (evacuate when directed or if you are uncomfortable). These terms are in the Santa Barbara County hazard education program [50], and drills are currently being tested in elementary schools in Montecito as part of hazard education [29, 30]. More suitable, during the 2018 event, many people did not comply with the evacuation order required by local stakeholders, and errors detected in the past were not taken into account by many actors. Understanding why people do not comply with evacuation orders concerning debris flows is key to knowing how to better communicate the risks in ways that may lead to better disaster preparedness and response. And even a slight delay in starting your evacuation will result in significantly longer travel times as traffic congestion worsens [50]. Therefore, the challenge still remains in identifying the exact timing and location where intense convective cells might develop (isolated or within a larger system). The 2018 event featured a north–south oriented atmospheric river with two moisture bands interacting with a closed low-pressure system. The main AR had moved southeast by the time of the debris flows. While the NCFR drove the high rain rates that produced the debris flow, the AR helped transport moisture into the area. Across the Santa Ynez and Topatopa Mountains, approximately 2–5+ inches of rain fell over a 2-day period. This value indicates a moderate storm for the region in terms of precipitation totals. However, the NCFR produced periods of intense rain. The 15-minute rains observed correspond to a 25–50 years event according to NOAA, while the FS model reported a 15 min total of 0.86 inches, a 100-year event [51].

6. Conclusions

The occurrence of the debris flows in Montecito in 2018 were not a surprise, but the magnitude and impacts of the flows were. Before the rains arrived, first responders were prepositioned, evacuations were implemented, a proclamation of emergency was executed at local scales and the operations center was staffed. However, despite technical or engineering advances, such sudden events occurred after extreme fires, and the susceptibility as thresholds indicating their probable occurrence could not be adapted to this remarkable situation. In addition, the Montecito event highlighted the need to develop more refined models that can be used in the field to accurately identify the risk and map debris flow inundation zones. Being able to map the hazard and related risk help convey this critical information to decision makers to implement mitigations and appropriate emergency measures, such as issuing an evacuation alert. Decision makers need to have confidence and be able to point to the science that what they are doing is proper and prudent. On the other hand, the studied catchments were investigated as a new example to detect a possible "cauliflower effect," and these new results confirm the influence of morphological conditions playing a key role in real postfire debris flow events. In the era of extremes, intense rainfall intensities are expected (at daily and not annual scales) but coupled with severe dry periods and severe fires, runoff concentration and violent hydrological responses could occur and surprise many people located at the outlet of small catchment. Thus, even if RuiCells© model should be more calibrated, we propose to assess such risk considering that catchments are totally burned, and the detection of "cauliflower effects" is finally not so theoretical. And to follow this study, we plan to measure this effect on the other postfire debris flows that occurred in California, with a set of 334 events of events.

Acknowledgements

The author addresses his sincere thanks to Anna Serra-Llobet, John Radke, and Matthias Kondolf (working at Berkeley), with whom he worked, and for the meeting organized the 6 February 2020 in Berkeley, with crisis managers and scientists, just before the international pandemic crisis. He also thanks Daniel Delahaye (University of Caen, France) and Patrice Langlois who create the RuiCells© model. He also thanks the institutions who financially supported this research: the France Berkeley Fund, for funds obtained in 2019; Avignon University, for support in 2022; and the Institut Universitaire de France (IUF), for grants over the period 2021–2026.

Author details

Johnny Douvinet
Avignon University, Avignon, France

*Address all correspondence to: johnny.douvinet@univ-avignon.fr

References

[1] Westerling A, Hidalgo H, Cayan D, Swenam T. Warming and earlier spring increase forest wildfire activity. Science. 2006;**313**:940-943. DOI: 10.1126/science.1128834

[2] Donat M, Lowry A, Alexander L, Ogorman P, Maher N. More extreme precipitation in the world's dry and wet regions. Nature Climate Change. 2016;**6**:508-513

[3] Swain D, Langenbrunner B, Neelin JD, Hall A. Increasing precipitation volatility in twenty-first-century California. Nature Climate Change. 2018;**8**:427-433

[4] Cannon S, DeGraff J. In: Sassa K, Canuti P, editors. The increasing wildfire and post-fire debris-flow threat in western USA, and implications for climate change, Landslides—Disaster Risk Reduction. Berlin, Heidelberg: Springer; 2008. pp. 177-190. DOI: 10.1007/978-3-540-69970-5_9

[5] Kean J, Staley D, Lancaster J, Rengers F, Swanson B, Coe J, et al. Inundation, flow dynamics, and damage in the 9 January 2018 Montecito Debris-flow event. California, USA: Opportunities and challenges for post-wildfire assessment. Geosphere. 2019;**15**:1140-1163. DOI: 10.1130/GES02048.1

[6] Cui Y, Cheng D, Chan D. Investigation of post-fire debris flows in Montecito. Geo-information. 2019;**8**:5. DOI: 10.3390/ijgi8010005

[7] Coe J, Kean J, Godt J, Baum R, Jones E, Gochis D, et al. New insights into debris-flow hazards from an extraordinary event in the Colorado Front Range. GSA Today. 2014;**24**(10):4-10. DOI: 10.1130/GSATG214A.1

[8] Takahashi T. Debris Flow: Mechanics, Prediction and Countermeasures. London, UK, Taylor and Francis; 2007. DOI: 10.1201/9780203946282

[9] Cannon S, Gartner J, Rupert M, Michael J, Staley D, Worstell B. Emergency Assessment of Postfire Debris-Flow Hazards for the 2009 Station Fire. San Gabriel Mountains, Southern California: U.S. Geological Survey Open-File Report; 2009. p. 27

[10] Costa J. Rheologic, geomorphic, and sedimentologic differentiation of water floods, hyper-concentrated flows and debris flows. In: Baker VR, Kochel RC, Patton PC, editors. Flood Geomorphology. Chichester: John Wiley & Sons; 1988. pp. 113-122

[11] Pierson T. Distinguishing between debris flows and flows from field evidence in small watersheds. U.S. Geological Survey Fact Sheet, 2004-3142. 2005

[12] Hungr O, McDougall S. Two numerical models for landslide dynamic analysis. Computers & Geosciences. 2009;**35**:978-992. DOI: 10.1016/j.cageo.2007.12.003

[13] Meyer G, Wells S. Fire-related sedimentation events on alluvial fans, Yellowstone National Park USA. Journal of Sedimentary Research. 1997;**67**(5):776-791

[14] Fuchs S, Kaitna R, Scheidl C, Hubl J. The application of the risk concept to debris flow hazards. 2008:120-129. Available from: http://wwwdsireusaorg/ (in English)

[15] Wondzell S, King J. Postfire erosional processes in the Pacific Northwest and

Rocky Mountain regions. Forest Ecology and Management. 2003;**178**:75-87

[16] Parise M, Cannon S. Wildfire impacts on the processes that generate debris flows in burned watersheds. Natural Hazards. 2012;**61**(1):217-227

[17] Moody J, Martin D, Cannon S. Post-wildfire erosion response in two geologic terrains in the western USA. Geomorphology. 2008;**95**:103-118

[18] Nyman P, Sheridan G, Smith H, Lane P. Evidence of debris flow occurrence after wildfire in upland catchments of south-east Australia. Geomorphology. 2011;**125**(3):383-401

[19] Parrett C. Fire-related debris flows in the Beaver Creek drainage, Lewis and Clark County, Montana. In: Subitzky S, editor. USGS Water Supply Paper 2330, Denver CO. 1987. pp. 57-67

[20] Cannon S, Gartner S. Runoff and erosion generated debris flows from recently burned basins. Geological Society of America Abstracts with Programs. 2005;**37**(7):35

[21] Wells W. The effects of fire on the generation of debris flows in southern California. In: Costa JE, Wieczorek GG, editors. Debris Flow/Avalanches: Process, Recognition, and Mitigation. Reviews in Engineering Geology. Vol. VII. Boulder, CO: Geological Society of America; 1987. pp. 105-114

[22] Booker F. Landscape and management response to wildfires in California: [MSc thesis], University of California, Berkeley. 1988

[23] Archetti R, Lamberti A. Assessment of risk due to debris flow events. Natural Hazards Review, American Society of Civil Engineers. 2003;**4**:115-125

[24] Gartner J, Cannon S, Santi P. Empirical models for predicting volumes of sediment deposited by debris flows and sediment laden floods in the transverse ranges of southern California. Engineering Geology. 2014;**176**:45-56

[25] Douvinet J, Van de Wiel M, Delahaye D, Cossart E. A flash flood hazard assessment in dry valleys (northern France) by cellular automata modeling. Natural Hazards. 2014;**75**(3):2905-2929. DOI: 10.1007/s11069-014-1470

[26] Delahaye D, Guermond Y, Langlois P. Spatial interaction in the runoff process. In: Proceedings of the 12th ECTG2001, European Colloquium on Theoretical and Quantitative Geography, Saint-Valéry-en-Caux, France. 2001

[27] Langlois P, Delahaye D. Ruicells, un automate cellulaire pour la simulation du ruissellement de surface. Revue Internationale de Géomatique. 2002;**12**:461-487

[28] Kean J, McGuire L, Rengers F, Smith J, Staley D. Amplification of post wildfire peak flow by debris. Geophysical Research Letters. 2016;**43**:8545-8553. DOI: 10.1002/2016GL069661

[29] Keller E, Adamaitis C, Alessio P, Anderson S, Goto E, Gray S, et al. Applications in geomorphology. Geomorphology. 2019;**366**:19. DOI: 10.1016/j.geomorph.2019.04.001

[30] Stock J, Dietrich W. Valley incision by debris flows: Evidence of a topographic signature. Water Resources Research. 2003;**39**(4):1089-1109. DOI: 10.1029/2001WR001057

[31] Kean J, Stanley D, Cannon S. In situ measurements of post-fire debris flows in southern California: Comparisons of the timing and magnitude of 24

debris-flow events with rainfall and soil moisture conditions. Journal of Geophysical Research. 2011;**116**:F04019. DOI: 10.1029/2011JF002005

[32] Santi P, deWolfe V, Higgins J, Cannon S, Gartner J. Sources of debris flow material in burned areas. Geomorphlogy. 2008;**96**:310-321

[33] BAER (U.S. Forest Service Burned Area Emergency Response). Thomas Fire Burn Area Reports, Online. 2018

[34] U.S. Geological Survey. Emergency assessment of post-fire debris-flow hazards, 2017 Thomas Fire, Online. 2018

[35] Restrepo P, Jorgensen D, Cannon S, Costa J, Laber J, Major J, et al. Joint NOAA/NWS/USGS prototype debris flow warning system for recently burned areas in southern California. Bulletin of the American Meteorological Society. 2008;**89**:1845-1851. DOI: 10.1175/2008BAMS2416.1

[36] NWS (National Weather Service). Flash flood and debris flow event, Montecito, California, January 9, 2018; Los Angeles, Oxnard, California, Online.

[37] Bessette-Kirton E, Kean J, Coe J, Rengers F, Staley D. An evaluation of debris-flow runout model accuracy and complexity in Montecito, California: Towards a framework for regional inundation-hazard forecasting. In: Kean JW, Coe JA, Santi PM, Guillen BK, editors. Debris-Flow Hazards Mitigation: Mechanics, Monitoring, Modeling, and Assessment, Proceedings of the 7th International Conference on Debris Flow Hazards Mitigation, Association of Environmental & Engineering Geologists Special Publication. Vol. 28. 2019. pp. 257-264

[38] Hamilton M, Serna J. Montecito Braced for Fire, But Mud Was a More Stealthy. Los Angeles Times: Deadly Threat; 2018

[39] Lai V, Tsai V, Lamb M, Ulzio T, Beer A. The seismic signature of debris flows: Flow mechanics and early warning at Montecito, California. Geophysical Research Letters. 2018, 2018;**45**:5528-5535. DOI: 10.1029/2018GL077683

[40] Langdon S, Johnson A, Sharma R. Debris flow syndrome: Injuries and outcomes after the Montecito Debris flow. The American Surgeon. 2019;**85**:1094-1098. DOI: 10.1177/000313481908501004

[41] Robert D, Niehaus I. The economic impacts of the Montecito Mudslides: First Assessment. 2018. Online

[42] Magnoli G. County estimates $46 million cost for Thomas fire, repairs. 2018

[43] Rodriguez-Iturbe I, Rinaldo A. Fractal River Basins, Chance and Self-organization. Cambridge: Cambridge University Press; 1997. p. 547. DOI: 10.1063/1.882305

[44] Fonstad M. Cellular automata as analysis and synthesis engines at the geomorphology-ecology interface. Geomorphology. 2006;**77**:217-234. DOI: 10.1016/j.geomorph.2006.01.006

[45] Palacios-Vélez O, Gandoy-Bernasconi W, Cuevas-Renaud B. Geometric analysis of surface runoff and the computation order of unit elements in distributed hydrological models. Journal of Hydrology. 1998;**211**:266-274

[46] Shreve R. Statistical law of stream numbers. Journal of Geology. 1966;**74**:17-37. DOI: 10.1086/627137

[47] Bak P. How Nature Works: The Science of Self-organized Criticality. New York: Springer-Verlag; 1996

[48] Murray A, Paola C. A cellular model of braided rivers. Nature. 1994;**371**:54-57

[49] Douvinet J, Serra-Llobet A, Radke J, Kondolf M. Quels enseignements tirer des coulées de débris post-incendie survenues le 9 janvier 2018 à Montecito (Californie, USA)? La Houille Blanche. 2020;**6**:25-35. DOI: 10.1051/lhb/2020052

[50] CAL Fire. Damage inspection database: California Department of Forestry and Fire Protection, Office of the State Fire Marshall, GIS file dated 26 April 2018. 2018

[51] Santa Barbara County. Santa Barbara County Department of Public Works rainfall data Online. 2018